Praise for *The Religion of Technology*

"For the last ten centuries or so in the West, Noble writes, technological visionaries have consistently spoken the language of Christian millenarianism. . . . As Noble demonstrates, this millenarian hope has taken a remarkable variety of forms. . . . There's much fascinating intellectual and social history in *The Religion of Technology*." —*The Boston Globe*

"Noble hits his stride when he reaches the twentieth century and starts flushing out the divine pretensions lurking in the fields of atomic weaponry, space exploration, artificial intelligence, and genetic engineering." —*The Village Voice*

"The first surprise in *The Religion of Technology* is that it succeeds so well in arguing that technology, not only throughout history but even into the present, has been at heart a religious endeavor. . . . Noble asserts that while technological development flourished when medieval monks deemed the practical arts and crafts a good way to pursue redemption, in the long run this has warped the type of technology developed. Too much of the technology produced since, he warns us, has been used for detrimental ends. . . . His point is that the millenarian strand of Christianity that has guided the development of technology and science now leads our technological society seriously astray." —*The Nation*

"Full of fascinating profiles of the architects of modern science. . . . Noble effectively gives lie to the myth that religion and modernization exist in a zero-sum game." —*Wired*

PENGUIN BOOKS

THE RELIGION OF TECHNOLOGY

David F. Noble is Professor of History at York University in Toronto. Currently the Hixon/Riggs Visiting Professor at Harvey Mudd College in Claremont, California, he has also taught at the Massachusetts Institute of Technology and Drexel University, and was a curator of modern technology at the Smithsonian Institution. His previous books include *America by Design: Science, Technology, and the Rise of Corporate Capitalism* and *Forces of Production: A Social History of Industrial Automation*.

THE RELIGION OF TECHNOLOGY

THE DIVINITY OF MAN AND
THE SPIRIT OF INVENTION

_____ DAVID F. NOBLE

PENGUIN BOOKS

FOR SOPHIE TENZER NOBLE

PENGUIN BOOKS
Published by the Penguin Group
Penguin Putnam Inc., 375 Hudson Street,
New York, New York 10014, U.S.A.
Penguin Books Ltd, 27 Wrights Lane,
London W8 5TZ, England
Penguin Books Australia Ltd,
Ringwood, Victoria, Australia
Penguin Books Canada Ltd, 10 Alcorn Avenue,
Toronto, Ontario, Canada M4V 3B2
Penguin Books (N.Z.) Ltd, 182–190 Wairau Road,
Auckland 10, New Zealand

Penguin Books Ltd, Registered Offices:
Harmondsworth, Middlesex, England

First published in the United States of America by Alfred A. Knopf, Inc. 1997
Published by arrangement with Alfred A. Knopf, Inc.
This edition with a new preface published in Penguin Books 1999

10 9 8 7 6 5 4 3 2

THE LIBRARY OF CONGRESS HAS CATALOGUED THE HARDCOVER AS FOLLOWS:
Noble, David F.
The religion of technology: the divinity of man and the spirit of
invention / by David F. Noble.—1st ed.
p.   cm.
Includes bibliographical references and index.
ISBN 0-679-42564-0 (hc.)
ISBN 0 14 02.7916 4 (pbk.)
1. Technology—Religious aspects—Christianity.   2. Religion and Science.
3. Presence of God.   I. Title.
BRII5.T4N63   1997
261.5´6—dc21   96–48019

Printed in the United States of America
Set in Sabon

This, then, is what the arts are concerned with, this is what they intend, namely, to restore within us the divine likeness.

—HUGH OF ST. VICTOR

We are agreed, my sons, that you are men. That means, as I think, that you are not animals on their hind legs, but mortal gods.

—FRANCIS BACON

Chicago physicist Richard Seed became a media sensation in the early days of 1998 when he proclaimed that he would proceed with experimentation on human cloning despite widespread public concern about the human and social implications of such research. His bold announcement sparked heated debate around the world and even prompted political efforts to legislate against scientific adventurism. Entirely ignored in all the clamor, however, was the religious rationale at the core of Seed's defiant declaration, which, in his view, placed his project above social concerns.

"God made man in his own image," Seed explained in a January 7th interview on National Public Radio's *Morning Edition*. "God intended for man to become one with God. We are going to become one with God. We are going to have almost as much knowledge and almost as much power as God. Cloning and the reprogramming of DNA is the first serious step in becoming 'one with God.'" With these words, Seed succinctly echoed the central refrain of the Religion of Technology, a faith that has fired the Western technological imagination for a thousand years. No doubt this otherworldly mythology has inspired great works in the service of humanity, but as Seed's announcement attests, it might well prove more a menace than a blessing. Unfortunately, Seed's religious message went unheard and unheeded, and thus failed to trigger any reflection about the deeper impulses underlying the scientific and technological enterprise. It did confirm, however, the central thesis of this book.

—David Noble
September, 1998

CONTENTS

ACKNOWLEDGMENTS

I would like to thank the following people for inspiring, instructing, and otherwise indulging me during the course of this effort: Phil Agre, Andrew Chaiken, Tracy Clemenger, Ted Daniels, Ashbel Green, Bert Hall, Sandra Harding, Stefan Helmreich, David Hess, Tom Hughes, Andrew Kimbrell, Jeff Kruse, Roger Launius, Kevin McGuire, Leonard Minsky, Mary Jo O'Connor, George Ovitt, J. D. Pipher, Lee Saegesser, Jan Sapp, Daniel Schenker, Don Stevens, and especially Douglas Noble, Mary Ann O'Connor, and Margaret Wertheim. I would also like gratefully to acknowledge the support of the Social Sciences and Humanities Research Council of Canada.

THE RELIGION OF TECHNOLOGY

TECHNOLOGY AND RELIGION

We in the West confront the close of the second Christian millennium much as we began it, in devout anticipation of doom and deliverance, only now our medieval expectations assume a more modern, technological expression. It is the aim of this book to demonstrate that the present enchantment with things technological—the very measure of modern enlightenment—is rooted in religious myths and ancient imaginings. Although today's technologists, in their sober pursuit of utility, power, and profit, seem to set society's standard for rationality, they are driven also by distant dreams, spiritual yearnings for supernatural redemption. However dazzling and daunting their display of worldly wisdom, their true inspiration lies elsewhere, in an enduring, other-worldly quest for transcendence and salvation.

With the approach of the new millennium, we are witness to two seemingly incompatible enthusiasms, on the one hand a widespread infatuation with technological advance and a confidence in the ultimate triumph of reason, on the other a resurgence of fundamentalist faith akin to a religious revival. The coincidence of these two developments appears strange, however, merely because we mistakenly suppose them to be opposite and opposing historical tendencies.

Ever since the eighteenth-century Enlightenment, which proclaimed the inevitable "secularization" of society, it has generally been assumed that the first of these tendencies would historically supersede the second, that the advance of scientific technology, with its rational rigors grounded in practical experience and material knowledge, signaled the demise of religious authority and enthusiasm based upon blind faith and superstition. Religion, presumably, belonged to the primitive past, secular science and technology to the mature future. Yet today we are seeing the simultaneous flourishing of both, not only side by side but hand in hand. While religious leaders promote their revival of spirit through an avid and accomplished use of the latest technological advances, scientists and technologists increasingly attest publicly to the value of their work in the pursuit of divine knowledge.[1]

Viewed from a larger historical perspective, there is nothing so peculiar about this contemporary coincidence, for the two tendencies have actually never been far apart. What we experience today is neither new nor odd but, rather, a continuation of a thousand-year-old Western tradition in which the advance of the useful arts was inspired by and grounded upon religious expectation. Only during the last century and a half or so has this tradition been temporarily interrupted—or, rather, obscured—by secularist polemic and ideology, which greatly exaggerated the allegedly fundamental conflict between science and religion. What we find today, therefore, is but a renewal and a reassertion of a much older historical tradition.

Some contemporary observers have argued, echoing generations of religious apologists, that the resurgence of religious expression testifies to the spiritual sterility of technological rationality, that religious belief is now being renewed as a necessary complement to instrumental reason because it provides the spiritual sustenance that technology lacks. There is perhaps some truth to this proposition, but it still presupposes the mistaken assumption of a basic opposition between these two phenomena and ignores what they have in common. For modern technology and modern faith are neither complements nor opposites, nor do they represent succeeding stages of human development. They are merged, and always have been, the technological enterprise being, at the same time, an essentially religious endeavor.

This is not meant in a merely metaphorical sense, to suggest that technology is similar to religion in that it evokes religious emotions of omnipotence, devotion, and awe, or that it has become a new (secular) religion in and of itself, with its own clerical caste, arcane rituals, and articles of faith. Rather, it is meant literally and historically, to indicate that modern technology and religion have evolved together and that, as a result, the technological enterprise has been and remains suffused with religious belief.

Perhaps nowhere is the intimate connection between religion and technology more manifest than in the United States, where an unrivaled popular enchantment with technological advance is matched by an equally earnest popular expectation of Jesus Christ's return. What has typically been ignored by most observers of these phenomena is that the two obsessions are often held by the same people, many among these being technologists themselves. If we look closely at some of the hallmark technological enterprises of our day, we see the devout not only in the ranks but at the helm. Religious preoccupations pervade the space program at every level, and constitute a major motivation behind extraterrestrial travel and exploration. Artificial Intelligence advocates wax eloquent about the possibilities of machine-based immortality and resurrection, and their disciples, the architects of virtual reality and cyberspace, exult in their expectation of God-like omnipresence and disembodied perfection. Genetic engineers imagine themselves divinely inspired participants in a new creation. All of these technological pioneers harbor deep-seated beliefs which are variations upon familiar religious themes.

Beyond the professed believers and those who employ explicitly religious language are countless others for whom the religious compulsion is largely unconscious, obscured by a secularized vocabulary but operative nevertheless. For they too are the inheritors and bearers of an enduring ideological tradition that has defined the dynamic Western technological enterprise since its inception. In the United States, for example, it must be remembered, industrialization and its corollary enthusiasm for technological advance emerged in the context of the religious revival of the Second Great Awakening. As historian Perry Miller once explained, "It was not only in the Revival that a doctrine of 'perfectionism' emerged. The revivalist mentality was sibling to the technological."[2]

But the link between religion and technology was not forged in the workshops and worship of the New World. Rather, the religious roots of modern technological enchantment extend a thousand years further back in the formation of Western consciousness, to the time when the useful arts first became implicated in the Christian project of redemption. The worldly means of survival were henceforth turned toward the other-worldly end of salvation, and over the next millennium, the heretofore most material and humble of human activities became increasingly invested with spiritual significance and a transcendent meaning—the recovery of mankind's lost divinity.

The legacy of the religion of technology is still with us, all of us. Like the technologists themselves, we routinely expect far more from our artificial contrivances than mere convenience, comfort, or even survival. We demand deliverance. This is apparent in our virtual obsession with technological development, in our extravagant anticipations of every new technical advance—however much each fails to deliver on its promise—and, most important, in our utter inability to think and act rationally about this presumably most rational of human endeavors.

Human beings have always constructed collective myths, in order to give coherence, a sense of meaning and control, to their shared experience. Myths guide and inspire us, and enable us to live in an ultimately uncontrollable and mysterious universe. But if our myths help us, they can also over time harm us, by blinding us to our real and urgent needs. This book describes the history of one such myth: the religion of technology. It is offered in the hope that we might learn to disabuse ourselves of the other-worldly dreams that lie at the heart of our technological enterprise, in order to begin to redirect our astonishing capabilities toward more worldly and humane ends.

PART I _____

TECHNOLOGY AND
TRANSCENDENCE

CHAPTER 1

THE DIVINE LIKENESS

The dynamic project of Western technology, the defining mark of modernity, is actually medieval in origin and spirit. A pattern of coherent, continuous, and cumulative advance in the useful arts, as opposed to a slow, haphazard accumulation of isolated specific inventions, emerged uniquely in the European Middle Ages. This unprecedented enterprise reflected a profound cultural shift, a departure from both classical and orthodox Christian belief, whereby humble activities heretofore disdained because of their association with manual labor, servitude, women, or worldliness came to be dignified and deemed worthy of elite attention and devotion. And this shift in the social status of the arts, if not the artisans, was rooted in an ideological innovation which invested the useful arts with a significance beyond mere utility. Technology had come to be identified with transcendence, implicated as never before in the Christian idea of redemption. The worldly means of survival were now redirected toward the other-worldly end of salvation. Thus the emergence of Western technology as a historic force and the emergence of the religion of technology were two sides of the same phenomenon.[1]

The other-worldly roots of the religion of technology were distinctly Christian. For Christianity alone blurred the distinction and bridged the divide between the human and the divine. Only here did salvation come to signify the restoration of mankind to its original God-likeness.

The uncompromisingly monotheistic Jews believed that they were the Chosen People of the one and only God, and blessed therefore with the burden of morality. For them it was always clear, however, what was God and what was man, a point pressed home in the Genesis story. In times of great trial, it is true, Jewish prophets resorted to rhetorical excesses in which their warrior Messiah, come to vanquish their enemies, deliver them from oppression, and rebuild Jerusalem, assumed supernatural dimensions. Thus, in the second century B.C., the prophet Daniel envisioned that the "Son of Man . . . came with the clouds of heaven" to establish "an everlasting dominion." And in the first century A.D. Apocalypses of Baruch and Ezra, the Messiah was endowed with miraculous powers enabling him to eliminate entirely strife, violence, want, and untimely death (though not death itself). But the Jews made little of such hopes and soon abandoned them altogether. Henceforth, as Norman Cohn noted, "it was no longer Jews but Christians who cherished and elaborated prophecies in the tradition of Daniel's dream."[2]

In its decidedly other-worldly reinterpretation of Old Testament prophecy, as the great sociologist of religion Max Weber long ago pointed out, Christian trinitarianism in effect revived Roman polytheism, this time giving men a place in the divine pantheon. "The incarnation of God presented men with the opportunity to participate significantly in God," Weber observed, "or as Irenaeus had already phrased it, 'enabled men to become gods.'" According to Augustine, the original Adam, having been created in God's image, was immortal, a distinctly divine characteristic forfeited with the Fall. Christ, the "Son of Man . . . come in the glory of his Father with his angels," was identified by Paul as the "last Adam," whose true divinity and immortality were revealed with the Resurrection, and was symbolically made accessible to his followers through the ritual of baptismal regeneration. Recalling the divine likeness of the first Adam, the advent of Christ promised the same destiny for a redeemed mankind. This was made explicit in the millenarian Book of Revelation, which prophesied a happy ending to the biblical story wherein all the righteous would regain their divinity in a succession of resurrections. "And God shall wipe away all tears from their eyes, and there shall be no more death."[3]

"By way of interpreting, therefore, through a definition the notion of Christianity," Gregory of Nyssa wrote in the fourth century, "we shall say that Christianity is the imitation of the divine nature. . . . For the first making of man was according to the imitation of God's likeness . . . and the promise of Christianity is that man will be brought back to the original happiness." For Christians, then, human efforts to recover Adamic perfection and imitate the life of Christ were one and the same: the pursuit of divinity. Through piety and asceticism, the saints strove to join the angels, and through their devoted exertions supposedly got at least halfway there. Under their aegis, the advance of the arts eventually became yet another means to the same exalted end.[4]

During the first Christian millennium, technology and transcendence belonged to entirely different realms. Even though both Christ and Paul had been artisans and many of the early adherents to the faith had come from the laboring classes, including women, the Church elite inherited a classical disdain for the useful arts. Moreover, after the fourth century, orthodox dogma, while recognizing the importance of such activities in easing the plight of fallen mankind, explicitly denied that they had any value as a means of redemption, which grace alone could provide.

"Quite apart from those supernatural arts of living in virtue and of reaching immortal beatitude which nothing but the grace of God which is in Christ can communicate to the sons of promise and heirs of the kingdom," Augustine, the chief author of Christian orthodoxy, wrote in *The City of God*, "there have been discovered and perfected, by the natural genius of man, innumerable arts and skills which minister not only to the necessities of life but also to human enjoyment." Augustine recognized the "astonishing achievements" that had taken place in cloth-making, navigation, architecture, agriculture, ceramics, medicine, weaponry and fortification, animal husbandry, and food preparation; in mathematics, astronomy, and philosophy; as well as in language, writing, music, theater, painting, and sculpture. But he emphasized again that "in saying this, of course, I am thinking only of the nature of the human mind as a glory of this mortal life, not of faith and the way of truth that leads to eternal life. . . . And, remember, all these favors taken together are but the fragmentary solace allowed us in a life condemned to misery."[5]

As Jacques Ellul, the theologian and philosopher of technology, argued, echoing Augustine, technology existed only for mankind in its fallen state, and had no significance beyond it. In its prelapsarian perfect state mankind had no need for such artifice, nor would it have in the renewal of that perfect state. In the Augustinian view, therefore, technology had nothing whatsoever to do with transcendence; indeed, it signified the denial of transcendence. Transcendence, the recovery of lost perfection, could be gained only by the grace of God alone. Moreover, those so blessed, said Augustine, would partake in a "universal knowledge" far beyond the ken of mere mortals. "Think how great, how beautiful, how certain, how unerring, how easily acquired this knowledge then will be. And what a body, too, we shall have, a body utterly subject to our spirit and one so kept alive by spirit that there will be no need of any other food."[6]

In the early Middle Ages, for reasons that remain obscure, the relationship between technology and transcendence began to change. Over time, technology came to be identified more closely with both lost perfection and the possibility of renewed perfection, and the advance of the arts took on new significance, not only as evidence of grace, but as a means of preparation for, and a sure sign of, imminent salvation. Historian Lynn White suggested that the changing attitude toward technology might have begun with the introduction of the heavy plow in the Frankish Empire. This major technological innovation radically reversed the relation between man and nature by making the capacity of a machine rather than human need the standard of land division: "Formerly he had been a part of nature; now he became an exploiter of nature." Shortly thereafter, around 830, a new form of calendar illustration began to appear among the Franks which highlighted this new attitude toward nature. Pictures of plow-ing, haying, and harvesting represented an active, coercive, dominating posture: "Man and nature are two things, and man is master." At the very same time, during the Carolingian age, there appeared what White described as "the earliest indication that men thought advancing technology to be an aspect of Christian virtue."[7]

In the Utrecht Psalter, illuminated near Rheims around 830, there is an illustration of Psalm 63 which gives technological advantage to those on the side of God. The army of the righteous confront a much larger army of the ungodly. "In each camp a sword is being sharpened conspicuously. The Evildoers are content to use an old-fashioned whetstone. The Godly, however, are employing the first crank recorded outside China to rotate the first grindstone known anywhere. Obviously the artist is telling us that technological advance is God's will."[8]

This ideologically innovative illustration was produced, according to White, "almost certainly by a Benedictine monk," an inference no doubt based upon the fact that Benedictine monks were not only prodigious scriptural illuminators but also earnest advocates of the arts in the service of spiritual ends. In the sixth century, Benedict of Nursia made the practical arts and manual labor in general vital elements of monastic devotion, alongside liturgical praise of God and the meditative reading of Scripture. Whatever its practical result—and monastic achievements in this regard were monumental—the true purpose of such effort was always, as George Ovitt has emphasized, the pursuit of perfection—"The monastic theorists favored manual labor but always as a means to a spiritual end"—and it was this overriding spiritual motivation that inspired such unprecedented performance. "It is one of the most amazing facts of Western cultural history," noted Ernst Benz, "that the striking acceleration and intensification of technological development in post-Carolingian Europe emanated from contemplative monasticism."[9]

It was under the imperial aegis of the Carolingians that the order of Benedictines first became hegemonic in Western Europe. Charlemagne imposed the Benedictine Rule on all religious houses in his realm, and his son Louis the Pious, an earnest advocate of useful innovation, was the original patron of the monastic-reform movement which was to sweep through Europe in the tenth and eleventh centuries. Under first imperial, and then feudal and papal auspices, the Benedictines eventually turned their religious devotion to the useful arts into a medieval industrial revolution, pioneering in the avid use of windmills, watermills, and new agricultural methods. In the process, the monastic elevation of technology as a means toward transcendent ends gained wider currency.

By investing them with a spiritual significance, the Benedictines lent a new dignity to the useful arts, which was reflected in Carolingian calendar illustrations and scriptural illumination. And as these illustrations indicated, this social elevation of the arts signified at the same time an ideological elevation of mankind above nature. In theological terms, this exalted stance vis-à-vis nature represented a forceful reassertion of an early core Christian belief in the possibility of mankind's recovery of its original God-likeness, the "image-likeness of man to God" from Genesis (1:26), which had been impaired by sin and forfeited with the Fall.[10]

In monasticism especially, the Christian pursuit of this renovation of the image-likeness of man became a collective rather than a merely individual compulsion—this stated objective appears in protection or exemption charters for monasteries. At the height of the Carolingian renaissance, which was profoundly influenced by monasticism, Alcuin, the chief of Charlemagne's famous palace school, used just such a notion to express his hope that a renewal of wisdom and knowledge had actually begun in Charlemagne's empire. And as Gerhart Ladner noted, thereafter "the idea of the reform of man to the image and likeness of God became the inspiration of all reform movements in . . . medieval Christianity."[11]

Moreover, the conception of image-likeness began to undergo a significant change in this period as well. The patristic view which had prevailed up to this point was that the divine image of man was purely spiritual in nature, located in the rational soul. Hence, the recovery of this original image entailed a necessary abandonment of the body, of matter. In the Carolingian age, particularly in the influential work of John Scotus Erigena, court philosopher to Charlemagne's grandson Charles the Bald, the notion of image-likeness for the first time incorporated the corporeal—the body and the external senses—

as a necessary correlate of reason and spirit. If the spirit required the corporeal, in this new view, the corporeal was in turn spiritualized, and matter became linked with the transcendent. It is likely that the Carolingian advances in, and heightened regard for, the useful arts reflected and reinforced such a transformed vision of the image-likeness of God to man. In the view of historian Ernst Benz, this belief ultimately became "one of the strongest impulses for man's technological development and realization. . . ." "Significantly," Benz wrote, "the founders of modern technology have felt that the justification of the most far-reaching aims of their technological efforts could be found in this very thought of the destiny of man as *imago dei* and his vocation as the fellow worker of God . . . to co-operate with God in the establishment of his Kingdom and . . . to share God's dominion over the earth." [12]

The new view of the useful arts, as distinct, dignified, divinely inspired, and of value for salvation, was first fully articulated in the ninth century, in the work of the Carolingian philosopher Erigena. By this time, the increasing attention given to the various technical arts and crafts by medieval observers had culminated in the coining of a new generic term, the "mechanical arts," to denote all of them collectively as a distinct category of human activity—the forerunner of the later terms "useful arts" and "technology." Augustine, for example, had no such vocabulary at his disposal, and referred the reader instead to the "innumerable arts and skills," "astonishing achievements," "contrivances," or to each particular craft in turn—cloth-making, navigation, etc. According to recent studies, the earliest known use of the term *artes mechanicae* to describe the arts collectively appears in Erigena's work, and thereafter, as interest in craftsmanship grew, the term came into common usage. Borrowed from Erigena, it was later used by Hugh of St. Victor in his enormously influential classification of knowledge. By the end of the twelfth century, the rubric had been fully absorbed into the mainstream of medieval thought and became the normal term for technological arts, used by such philosophers as Abelard, Duns Scotus, Bonaventura, Albertus Magnus, and Raymond Lully. [13]

Erigena coined the term "mechanical arts" in his commentary

on Martianus Capella's fifth-century work *The Marriage of Philology and Mercury*. He not only recognized the various useful arts as constituting a distinct class of activities but also, in stark contrast to Capella, accorded them an unprecedented status equal to that of the seven liberal arts. In Capella's work, Mercury gives his new bride the gift of those seven arts—Grammar, Dialectic, Rhetoric, Geometry, Arithmetic, Astronomy, and Harmony—each represented in a performance by a maiden. Capella pointedly omits from this nuptial performance the two mechanical disciplines Medicine and Architecture, because of their "baseness" and "unworthiness." "Since these ladies are concerned with mortal subjects and their skill lies in mundane matters, and they have nothing in common with the celestial deities," wrote Capella, "it will not be inappropriate to disdain and reject them."[14]

In a radical departure from tradition, Erigena rewrote Capella's allegory to include the heretofore disdained mechanical arts. In his new version, the bride Philology, after receiving Mercury's gift of the liberal arts, gives him in return the parallel gift of seven mechanical arts, including Medicine and Architecture. Thus the mechanical arts, though not actually included among the liberal arts, are nevertheless represented as having equal significance.

In giving the mechanical arts such status, Erigena implied that, even if no doubt concerned with "mundane matters," they nevertheless shared something "in common with the celestial deities." There was a connection between the mundane and the celestial, between technology and transcendence. Just as he insisted upon the significance of the corporeal, physical element in the image-likeness of man to God, in service to the spiritual, so he likewise insisted upon the significance of the arts in the restoration of that image, in the service of salvation. Departing from the Augustinian view, Erigena argued that the useful arts were indeed part of mankind's original endowment, his God-like image, rather than merely a necessary product of his fallen state. Thus the mechanical arts rightfully had an honored place in divine creation. Erigena insisted that knowledge of the arts was innate in man, an aspect of his initial endowment, but that it had become obscured by sin since the Fall of Man, and was now but a dim vestige of its original perfection. He believed, however, that, through practical effort and study, mankind's prelapsarian powers could be at least partially recovered and could contribute, in the process, to the restoration of perfection. In other words, Erigena invested the arts with spiritual significance, as elements of man's God-likeness, and identified them as vehicles of redemption. As one scholar summarized Erigena's thought, "In pursuing the study of the arts . . . one progresses in perfection since the arts are innate in man. Knowledge of them has been clouded by the Fall. Their recovery by study helps to restore man to his pristine state." [15]

The arts, Erigena wrote, are "man's links with the Divine, their cultivation a means to salvation." He declared that "every natural art is found materially in human nature," and argued that "it follows that all men by nature possess natural arts, but, because, on account of the punishment for the sin of the first man, they are obscured in the souls of men and are sunk in a profound ignorance, in teaching we do nothing but recall to our present understanding the same arts which are stored deep in our memory." Erigena's boldly innovative and spiritually promising reconceptualization of the arts signaled a turning point in the ideological history of technology. As one Erigena scholar noted, "It would be difficult to over-estimate the significance of this development. The new emphasis on the place of the arts in Christian education must be seen as one of the chief factors animating the ninth century's intense interest in the arts." This new "Christianization of the arts" for the first time gave the means of mortal survival a crucial role in the realization of immortal salvation. [16]

Legend has it that late in life Erigena became an abbot of a Benedictine monastery in England. Whether this is true or not, there is little doubt that Erigena's new conception of the useful arts was sustained by the monastic community that had inspired it. His use of the term "mechanical arts" reappears, for example, in a later monastic commentary on Capella by Remigius of Auxerre. His notion that the mechanical arts had been divinely inspired was illustrated in a new iconography of the creator God as master craftsman, which first appeared at the end of the tenth century in Winchester, an important site of Carolingian-inspired monastic reform. Here the monastic illuminator of a gospel book made what Lynn White described as a "great innovation," for the first time portraying the hand of God holding scales, a carpenter's square, and a pair of compasses—which later became the medieval and Renaissance symbol of the engineer. Around the same time, the Benedictines of Winchester Cathedral installed the first giant organ, the most complex machine known before the invention of the mechanical clock.[17]

But it was in the "mechanism-minded" world of the twelfth century that the new exalted, spiritualized view of the useful arts truly became the norm, especially among the innovative Cistercians and other Benedictines. The proliferation of new devices—watermills, windmills, mechanisms for metal-forging and ore-crushing, the mechanical clock, eyeglasses, the springwheel—both reflected and reinforced this new sensibility.[18]

In the first half of this century, the monastic technical tradition found its greatest written expression in a technical treatise by the German Benedictine Theophilus. A skilled metallurgist and general craftsman as well as a monk, Theophilus was "the first man in all history to record in words anything approaching circumstantial detail of a technique based on his own experience," according to metallurgist and historian of technology Cyril Stanley Smith. Theophilus's book, De Diversis Artibus, was "a religiously motivated codification of all the skills available for the embellishment of a church," including machine design, metal-casting, enameling, painting, glass-making, wire-drawing, and tinning. His reverence for such crafts was notable, especially in a world in which most craftsmen were either slaves or domanial serfs. Goldsmiths and ironsmiths had sometimes enjoyed a relatively privileged status because of the honorific rather than productive value of their work, such as the making of coins, jewels, and weapons. Here the arts were exalted because of their association with spiritual devotion. For Theophilus too, George Ovitt noted, "spiritual goals were primary. . . . Practical matters were pursued for the glory of God and the perfection of self."[19]

In the manner of Theophilus, the abbot Arnold of Bonneval marveled at the technical innovations introduced with the rebuilding of Clairvaux, the great mother abbey of the Cistercians, devoting detailed attention especially to the water-powered machinery for milling, fulling, tanning, and blacksmithing, which constituted what has been described as a veritable medieval industrial revolution. Another monastic observer of Clairvaux described an automatic flour-sifter and fulling mill, and, awed by the "abstract power of water flowing through the abbey seeking every task," thanked God for such labor-saving technology. The monastic mechanization of the crafts, as well as major construction projects such as the building of churches and aqueducts, had indeed become, and was clearly recognized as, "holy labor."[20]

The twelfth-century spiritualization and hence elevation of such practical activity was fully acknowledged and powerfully reinforced in the extremely influential work of the Augustinian canon Hugh of St. Victor. In his innovative classification of knowledge, the *Didascalicon,* Hugh gave "unprecedented psychic dignity and speculative interest to the mechanic arts." Greatly inspired by Erigena's commentary on Capella, Hugh borrowed Erigena's rubric "mechanical arts" as "a generic term for all crafts." Moreover, he elaborated on Erigena's creative recasting of Capella's allegory by specifying in detail the seven mechanical arts offered by Philology to Mercury in return for the seven liberal arts. These included cloth-making, armaments and building, commerce, agriculture, hunting and food preparation, medicine, and theatrics.[21]

Inspired by Erigena's ideas, Hugh likewise "linked the mechanical as well as the liberal arts directly to salvation and the restoration of fallen man." Although as an Augustinian Hugh identified technology exclusively with the fallen world (and with the first act of fallen man, the making of clothing), he nevertheless maintained, in a marked departure from Augustine, that the useful arts constituted a means of recovering mankind's perfection, his original divine image. Following Erigena, Hugh believed that this prelapsarian perfection was not solely spiritual, as Augustine had argued, but physical as well. Hence, he argued that "the work of restoration included the repair of man's physical life" as well as the spiritual. For Hugh, according to medievalist Elspeth Whitney, "the mechanical arts supply all the remedies for our physical weakness, a result of the Fall, and, like the other branches of knowledge, are ultimately subsumed under the religious task of restoring our true, prelapsarian nature." Hence, "through its relationship to man's final end, the pursuit of the mechanical arts acquired religious and moral sanction." "This, then, is what the arts are concerned with," wrote Hugh of St. Victor, "this is what they intend, namely, to restore within us the divine likeness."[22]

With Hugh the monastic reconception of the useful arts was fully articulated as a means of reunion with God, a theme sustained in the thirteenth century by Michael Scot, who held that "the primary purpose of the human sciences is to restore fallen man to his prelapsarian position," and by the Franciscan friar Bonaventura, who likewise "sanctified the mechanical arts and placed them in the context of knowledge whose source and goal is the light of God." Such work—by a canon, a layman, and a mendicant friar—not only further ratified the moral virtuousness of the useful arts but also helped to spread such monastic ideas beyond the cloister, fostering in Europe a unique emotional commitment to machinery, grounded upon an "acceptance of mechanisms as aids to the spiritual life."[23]

CHAPTER 2

MILLENNIUM: THE PROMISE OF PERFECTION

While successive generations of monks dedicated themselves anew to the recovery of mankind's divinity, their pious efforts lacked any tangible record of cumulative accomplishment. With the identification of the advancement of the useful arts as a means toward that exalted end, however, their striving gained concrete expression, and hence enduring evidence of their progress toward perfection. The development of technology now gave some assurance that mankind was indeed on the road to recovery. Accordingly, technological invention was duly incorporated into biblical commentary and thus Christian history.

At the same time, beginning in the middle of the twelfth century, there emerged from within the monastic world a radically renewed millenarian conception of Christian history, a dynamic and teleological sense of time which would profoundly excite Christian expectation and accelerate the technological development that was now bound up with it. For Augustine, historical time, the tiresome and tearful tenure of fallen man, was homogeneous and unchanging; the resurrection of Christ was a sign of promise, to be sure, but history offered no other indication of movement toward a restoration of perfection. Only God knew the agenda, which was hidden from man; if any correlation existed between human events and divine purpose, it could never be known. The new millenarian mentality changed all this. An elite revitalization and reinterpretation of early Christian belief, it situated the process of recovery in the context of human history and redefined it as an active and conscious pursuit rather than a merely passive and blind expectation. Moreover, it broke the divine code about human destiny, about the true relationship between the temporal and the transcendent, thereby offering both evidence of past progress and guidance for the future. The recovery of mankind's divine likeness, the transcendent trajectory of Christianity, thus now became at the same time an immanent historical project. As a result, the pursuit of renewed perfection—through myriad means which now included the advancement of the arts—gained coherence, confidence, a sense of mission, and momentum. This new historicized millenarianism was to have enormous and enduring influence upon the European psyche, and it encouraged as never before the ideological wedding of technology and transcendence. Technology now became at the same time eschatology.[1]

The Christian notion of the millennium is based upon the prophecy of the Book of Revelation, the last book of the Bible (known also as the Apocalypse of St. John), which was itself derived from ancient Hebrew prophecy. In his vision, John of Patmos foretells a thousand-year reign on earth of the returned Messiah, Christ, together with an elite corps of the saintly elect. In effect, this last book of the Bible is a return to the first book, Genesis, only now with happy ending. Here the fate of the Fall is reversed, the curse is lifted, and a redeemed mankind is permitted to return to paradise, eat from the tree of life, and regain Adam's original perfection, immortality, and godliness.

> Remember therefore from whence thou art fallen, and repent, and do the first works. . . . To him that overcometh will I give to eat of the tree of life, which is in the midst of the paradise of God. . . .

And I looked, and, lo, a Lamb stood on mount Sion, and with him an hundred and forty four thousand, having his Father's name written on their foreheads. . . . And they sung as it were a new song before the throne. . . . And he that sat upon the throne said, Behold, I make all things new. . . .

And he said unto me, It is done. I am Alpha and Omega, the beginning and the end. I will give unto him that he is athirst of the fountain of the water of life freely. He that overcometh shall inherit all things; and I will be his God, and he shall be my son. . . .

And he showed me a pure river of water of life, clear as crystal, proceeding out of the throne of God and of the Lamb. In the midst of the street of it, and on either side of the river, was there the tree of life. . . . And there shall be no more curse. . . .

Blessed and holy is he that hath part in the first resurrection. . . . They shall be priests of God and of Christ, and shall reign with him a thousand years.[2]

Millenarianism is, in essence, the expectation that the end of the world is near and that, accordingly, a new earthly paradise is at hand. In the early centuries of the Christian era, there were myriad millenarian voices heralding the imminent advent of the Kingdom of God, which drew their inspiration from biblical prophecy and mystical vision. But these voices were soon marginalized by the clerical caste, which embodied the power and authority of the Great Church. In the view of this emergent elite, the millennium had already begun with the establishment of the Church and they were the earthly saints. In their eyes, belief in a millennium yet to come was subversive, because it suggested that the Kingdom of God had not yet arrived but belonged to a future time beyond the Church. Thus, whereas at the end of the second century Bishop Irenaeus of Lyons could readily sanction and personally endorse millenarian expectations, his writings on such matters were eventually suppressed; in 431, the Council of Ephesus formally condemned millenarian belief as heresy.[3]

Despite official condemnation, belief in a future millennium continued to flourish, mostly as an expression of popular desperation and dissent. The medieval ecclesiastical elite neither offered nor harbored hope of an earthly paradise beyond the Church. In the high Middle Ages, however, in the wake of religious revival, a rigorist Church-reform movement, the Crusades, and renewed external threats to Christendom, millenarianism regained a degree of elite respectability, especially among the new religious orders, which made use of apocalyptic mythology to validate their identity and destiny, and thereby magnify their significance.[4]

The founding prophet of this renewed expectation was a Cistercian abbot from Calabria, Joachim of Fiore. In pursuit of the most perfect form of monasticism, this ardent, rigorously ascetic monastic reformer ultimately left the Cistercian order to establish his own monastery in Fiore, which he named after St. John. Joachim had been greatly influenced by the monastic and Church-reform movements, by the Crusades, and by the seemingly apocalyptic conflicts between popes and emperors, Christianity and Islam. For Joachim, Antichrist had appeared in the human form in Saladin, who conquered Jerusalem in 1187, signifying that the millennium was at hand. In his eyes, the reformed monks constituted the saintly vanguard of redeemed mankind, prepared not to challenge but to defend the established order of Christendom.

Inspired by a vision while reading the Book of Revelation, Joachim formulated what has been described as the "most influential prophetic system known to Europe until Marxism," which "ignited the greatest spiritual revolution of the Middle Ages." In his vision, Joachim wrote, the millenarian meaning of history, God's plan for humanity, was revealed to him. He taught that the divinely predetermined structure of history could be known through study of biblical prophecy, particularly the prophecy of St. John. In this light, there was a discernible pattern to history; it had momentum, direction, and meaning based upon the final events toward which it moved—the millennial reunification of man with God. In his *Exposition on the Apocalypse,* Joachim declared that the prophecy of St. John was "the key of things past, the knowledge of things to come; the opening of what is sealed; the uncovering of what is hidden." Through his new insight into the meaning of biblical prophecy, he claimed to be able not only to interpret the significance of human events up to that time but, more important, to read the signs of, and thus predict, events yet to come. Armed with such foreknowledge, which included an anticipation of their own appointed role, the elect needed no longer to just passively await the millennium; they could now actively work to bring it about.[5]

Joachim described the historical movement toward the millennium as a succession of three stages, each representing an element of the trinity. The first stage, that of the Father, was the *ordo conjugatorum*, initiated by Adam and symbolized by the family and the married state. The second, that of the Son, was the *ordo clericum*, initiated by Christ and embodied by the priesthood. The third and final stage of history, that of the Holy Spirit, was the *ordo monachorum*, initiated by St. Benedict and represented by the monk. This third stage, a period of transition which Joachim believed was in its final phase of millennial preparation, was an age marked by the appearance of the *viri spirituales*, the spiritual men who constituted the saintly vanguard of redeemed humanity. For Fiore, these were "the order of monks to whom the last great times are given." Through spiritual contemplation and preaching, they would bring about a general spiritual illumination and release mankind from its misery.[6]

Joachim, who became the apocalyptic consultant for three popes as well as the most powerful rulers of his age, believed that the millennium, anticipated in the devotion of his monastic disciples, was due to arrive in the year 1260. Soon after his death in 1202, however, the mantle of the third stage was claimed by a new breed of spiritual men, the mendicant friars. The Franciscans, especially the more radical or "spiritual" followers of St. Francis of Assisi, emphasized their transitional role as preachers in the world rather than mere contemplatives in the cloister. The millenarian prophecy of Joachim of Fiore provided these reformers with an understanding of their own historical mission in the world; they avidly edited and commented on his writings, which appeared to confirm their pre-eminent and predestined role in the pursuit of the millennium. Thus, despite continued official condemnation, which still put even elite millenarians in jeopardy, the prophetic teachings of Joachim of Fiore steadily became part of the "common stock of European social mythology."[7]

The mendicants were themselves succeeded as the bearers of the third stage by centuries of self-anointed successors, and each in turn added new dimensions to millenarian preparation. The Franciscans themselves, having emphasized evangelizing over contemplation, also acknowledged another means of millenarian anticipation: the advancement of the arts. By the thirteenth century, this millenarian inspiration behind technological development was already being anonymously represented in the work of countless cathedral-builders, the most advanced artisans of their time, whose silent stone images suggest a preoccupation with divine judgment and the world's end. Their efforts to improve technical skills were not conceived as a means of improving the condition of man within the present order of things, Arnold Pacey noted. "Rather, they were reaching forward to meet an eternal order, a new Jerusalem, which the cathedral itself symbolized."[8]

At the same time, some of the more radical Franciscans began to give voice to this new artistic mentality, none more forcefully than Roger Bacon. Having inherited the new medieval view of technology as a means of recovering mankind's original perfection, Bacon now placed it in the context of millenarian prophecy, prediction, and promise. If Bacon, following Erigena and Hugh of St. Victor, perceived the advance of the arts as a means of restoring humanity's lost divinity, he now saw it at the same time, following Joachim of Fiore, as a means of anticipating and preparing for the kingdom to come, and as a sure sign in and of itself that that kingdom was at hand.

Joachimite millenarianism linked the events of history with the end of history. Roger Bacon, the legendary Franciscan scholar who studied and taught at the universities of Oxford and Paris during the thirteenth century, was steeped in this new medieval tradition. Typically portrayed as a farsighted visionary of modern technological progress, Bacon was actually moored in his own millenarian milieu. If he recognized the practical potential of natural philosophy, urged greater development of the arts, and envisioned such modern inventions as self-powered cars, boats, submarines, and airplanes, he did so only with reference to the end-times, which he believed were already at hand.[9]

"All wise men believe that we are not far removed from the times of Antichrist," wrote Bacon, who was greatly influenced by the legacy of Joachim of Fiore. Bacon cited Joachim's authority in suggesting that the contemporary Tartar invasions signaled the arrival of Antichrist. Like Joachim, Bacon was himself an ascetic reformer who condemned the decadence of the world, the corruption of the Church, and the quarrels between the religious orders, and viewed them also as signs of the coming of Antichrist. He urged his fellow Franciscans and the Church to study Joachimite prophecy in order to be forewarned about history's final events; he continually referred to the 144,000 elect of the Book of Revelation who would lead the battle against Satan, was fixated by the specter of the Antichrist, and invoked the idea of an angelic pope as symbol of Joachim's third stage.[10]

It was in this apocalyptic spirit that Bacon counseled the pope to develop the useful arts. He warned that "Antichrist will use these means freely and effectively, in order that he may crush and confound the power of this world," and urgently advised that "the Church should consider the employment of these inventions . . . because of future perils in the times of Antichrist which with the grace of God it would be easy to meet, if prelates and princes promoted study and investigated the secrets of nature and art."[11]

At the same time, Bacon believed, following the tradition of Erigena and Hugh of St. Victor, that the arts were the birthright of the "sons of Adam," that they had once been fully known while mankind still reflected the image of God, that they had been lost because of sin but had already been partially regained, and that they might yet be fully restored, as part of the recovery of original perfection, through diligent and devout effort. In his Opus Majus, Bacon declared that "philosophy in its perfection" had initially been granted to man by God, in particular to "the saints at the beginning." He identified the causes of error in human knowledge with the Fall: "Owing to original sin and the particular sins of the individual, parts of the image have been damaged, for reason is blind, memory weak, and the will depraved." And yet he maintained that "truth gains strength and will do so until the day of Antichrist." Philosophy, Bacon wrote, "is merely the unfolding of the divine wisdom by learning and art," the "whole aim" of which "is that the Creator may be known through the knowledge of the creature." His manuscript ends with the promise of mankind's renewed divinity through reunification with God: "from participation in God and Christ we become one with him and one in Christ and are gods. . . . And what more can a man seek in this life?"[12]

Though Bacon emphasized the usefulness of knowledge, his notion of utility was decidedly other-worldly. He defiantly declared his contempt of the world and concerned himself instead with the "things which lead to felicity in the next life." For Roger Bacon, the advance of technology was doubly dedicated to the transcendent end of salvation: on the one hand, as the means of recovering the knowledge of nature which was part of mankind's divine inheritance, its original image-likeness to God, and, on the other, as the means of triumph over Antichrist in anticipation of the millennium. If the monastics had elevated the useful arts as a means of restoring their own original perfection, now mendicants like Bacon dignified them further by proclaiming their providential purpose in the historical pursuit of this perfection, as preparation for the millenarian redemption of humanity.[13]

In the thirteenth and fourteenth centuries, other radical Franciscan advocates of the arts followed Bacon's lead, among them the famed triumvirate of Catalan science, Raymond Lully, Arnau de Villanova, and John of Rupescissa. Lully, a Franciscan tertiary, was a practicing physician as well as an astrologer and was renowned for his knowledge of chemistry and metallurgy. Like Bacon, he was also steeped in the prophetic tradition. In his *Ars Magnus*, he claimed that his "Art," which he hoped could be used to convert the Arabs to Christianity, came to him through divine illumination. He wrote voluminous commentary on Joachim as well as his own prophecies about the coming of the Antichrist, drawing upon the biblical prophecies of Ezekiel, Daniel, and especially the Apocalypse of St. John. Also like Bacon, Arnau de Villanova was very close to the spiritual Franciscan movement as well as a strong advocate of natural science. He was known for medical and alchemical works in which criticism of the Church was "combined with Joachimite ideas of a speedy end of the world and coming of the Antichrist."[14]

John of Rupescissa, another Franciscan tertiary, has been recognized as the true founder of medical chemistry, whose work, especially on the distillation and the medical efficacy of alcohol, signaled a shift in chemistry from qualitative to quantitative methods of investigation. The bulk of his effort, however, was devoted to theological and particularly prophetic writing. He was familiar to his contemporaries for his apocalyptic preaching, because of which he spent a considerable part of his life in prison. As evangelists and missionaries in the world, the Joachimite Franciscans carried the millenarian message beyond the cloister. At the same time, in their preaching and writing they formulated what would become an enormously influential and enduring eschatololgy of technology, a perception of the advancing useful arts as at once an approximate anticipation of, an apocalyptic sign of, and practical preparation for the prophesied restoration of perfection.[15]

If some Franciscans promoted the arts directly, like Bacon and his Catalan successors, most did so indirectly, through their primary evangelical mission, the conversion of all races to Christianity. According to biblical and Joachimite prophecy, particularly the Book of Revelation, such worldwide conversion was a necessary precondition for, and unmistakable indication of, the coming of the millennium. "God has been calling all the peoples of the earth to hasten to prepare themselves to enter and to enjoy that everlasting feast that will be endless," wrote the sixteenth-century Franciscan missionary to the New World Geronimo de Mendieta. "This vocation of God shall not cease until the number of the predestined is reached, which according to the vision of Saint John must include all nations, all languages, and all peoples."[16]

The evangelical effort to extend the reach of Christianity in accordance with its universalist claims and eschatological expectations, moreover, encouraged exploration, and thereby advanced the arts upon which such exploration depended, including geography, astronomy, and navigation, as well as shipbuilding, metallurgy, and, of course, weaponry. "The striving to fulfill prophecy on a cosmic or global scale was a major stimulus to travel and discovery, from the early Franciscan missions into Asia to Columbus' Enterprise of the Indies," historian Pauline Moffitt Watts has noted. And this "apocalypticism" of explorers, she stressed, particularly on the part of Columbus, "must be recognized as inseparable" from their "geography and cosmology," because it both shaped their scientific understanding and inspired their technological accomplishment.[17]

The Age of Discovery really began in the middle of the thirteenth century, when mendicant friars (and merchants) traveled overland to Central and East Asia. The land route to the Far East was opened at that time by the Franciscan friar Giovanni da Pian del Carpini. One of the first writers "to integrate into an apocalyptic scheme the possibility of converting all the peoples of Asia, that is, all the rest of the known world," was John of Rupescissa. Rupescissa prophesied that the Tartar dynasty of Genghis Khan would be converted to Christianity (along with the Jews) and would thereafter join forces with Christians for the final defeat of Islam. Such evangelical expectations were short-lived, however: by the middle of the fourteenth century, "Islam had won the soul of Tartary" and "the land route to Asia was closed."[18]

Early-fourteenth-century Portuguese explorations of Africa initiated the oceanic phase of the Age of Discovery, and inspired evangelical hopes of an alternative sea route to Asia. These were ultimately fulfilled by the messianic mariner and "chiliastic crusader" Christopher Columbus, who believed himself divinely sent to open up a new way for the friars to fulfill the prophecies of the apocalypse, to convert the heathen, and to hasten the millennium.

The image of Columbus which emerges from most historical accounts is that of the intrepid, modern-minded mariner armed with new scientific understanding and rational methods, as well as a lifetime of practical experience as a navigator, mapmaker, and sailor, which enabled him to defy and overcome the ignorance and superstition of his contemporaries. There is no doubt about his technical prowess. "When I was very young I went to sea to sail and I continue to do it today," Columbus wrote in the preface to his *Book of Prophecies*. Over the years God "has bestowed the marine arts upon me in abundance and that which is necessary to me from astrology, geometry, and arithmetic. He has given me adequate inventiveness in my soul and capable hands." But, in Columbus's own view, these technical capabilities alone did not suffice to inspire him to undertake, or enable him to accomplish, the great deeds for which he is known. Rather, such skills were joined to another kind of endowment, without which they would have produced nothing.[19]

"This [sailor's] art predisposes one who follows it towards the desire to know the secrets of the world," Columbus explained, which led him in his life to seek and gain an understanding of prophecy and his appointed role in it. "Reason, mathematics and mappaemundi were of no use to me in the execution of the enterprise of the Indies," he insisted, without such divine inspiration and guidance; his achievement was, in reality, "a very evident miracle."[20]

If his voyages carried the world into the modern age, Columbus's own mentality reflected the medieval millenarian expectations of fifteenth-century Spain. In this spiritually charged setting, the Spanish monarchs assumed the mantle of Joachimite messiah-emperors of the third age, leading the righteous into the millennium. According to the Franciscan friar Geronimo de Mendieta, for example, "the Spanish race under the leadership of her 'blessed kings' had been chosen to undertake the final conversion of the Jews, the Moslems, and the Gentiles . . . , an event which foreshadowed the rapidly approaching end of the world." (Both the final defeat of the Moors in Granada in 1492 and the forced conversion, or expulsion, of the Jews that same year were perceived in this light.)[21]

That Christopher Columbus dedicated his life to this evangelical challenge is obvious from the first entry in his journal of his 1492 voyage. "Your Highnesses, as Catholic Christians, and princes who love and promote the Christian faith, and are enemies of the doctrine of Mahomet, and of all idolatry and heresy, determined to send me, Christopher Columbus, to the above mentioned countries of India, to see the said princes, people and territories, and to learn their disposition, and the proper method of converting them to our holy faith; and, furthermore, directed that I should not proceed by land to the East, as is customary, but by a Westerly route, in which direction we have hitherto no certain evidence that anyone has gone. So, after having expelled the Jews from your dominions, your Highnesses . . . ordered me to proceed."[22]

Very much a product of his times and culture, the great explorer was spiritually and intellectually well prepared for this challenge. According to his son Ferdinand, he lived a pious and ascetic life rigorous enough to "have been taken for a member of a religious order." His closest companions were monks and friars, especially Franciscans, with whom he associated and identified. He spent considerable time and prepared for his expeditions in monasteries. After his second voyage, he walked the streets of Seville and Cádiz dressed in the sackcloth of a penitent, and appeared indistinguishable from his Franciscan friends. On his deathbed, he took the habit of a Franciscan tertiary, and he was buried in a Carthusian monastery.[23]

Intellectually, Columbus was enormously influenced by the medieval millenarian and scientific traditions, primarily through the work of Cardinal Pierre d'Ailly. Neither an innovative thinker nor a Franciscan himself, d'Ailly was nevertheless an expert explicator of medieval masters, whose work he described in detail in *Imago Mundi*. Published in 1410, this compendium of ancient and medieval cosmology and geography circulated extensively in Western Europe throughout the fifteenth century. Here d'Ailly combined studies of geography, astronomy, meteorology, and calendar reform with an earnest advocacy of natural science in general. He shared with Roger Bacon, from whose work he borrowed most heavily, an equally ardent interest in the use of astrology as a guide to interpreting prophecy, defended Bacon's advice to Pope Clement IV on this subject, and was himself "especially interested in the coming of Antichrist and the end of the world, both of which he believed might be astrologically conjectured."[24]

D'Ailly was Columbus's chief source, for both his scientific geography and his apocalyptic outlook. Columbus carefully read and annotated the *Imago Mundi,* and used the knowledge it provided both to guide him in his voyages and to situate them in the divine millennial scheme. Through d'Ailly, Columbus became acquainted with the writings of Roger Bacon and the prophecies of Joachim of Fiore, which shaped his own reading of events.

Columbus saw himself as a "divinely inspired fulfiller of prophecy." He was firmly convinced that the world would end in about a century and a half, based upon calculations by d'Ailly, and that in the meantime all prophecies had to be fulfilled, including the conversion of all peoples and the recovery of Mount Zion (Jerusalem). According to his son, Columbus's given name, Christoferens (Christ-bearer), symbolized by the dove of the Holy Spirit, signified that, in the manner of his namesake St. Christopher, he had been chosen to carry the Christ child across the waters. Columbus himself later dubbed his effort the "enterprise of Jerusalem," and insisted that his voyages to the New World must be capped by a crusade to recapture the Holy Land and rebuild the Temple on Mount Zion. Above all else, Columbus believed himself guided by divine prophecy, which was the secret of his sublime confidence. "Who would doubt . . . this light, which comforted me with its rays of marvelous clarity . . . and urged me onward with great haste continuously without a moment's pause," he wrote his patrons. He proclaimed himself to be the Joachimite Messiah sent by God to prepare the world for its glorious end and renewed beginning. He was assured in this by a prophecy of Arnau de Villanova, which he mistakenly attributed to Joachim, that "he who will restore the ark of Zion will come from Spain."[25]

In his unfinished *Book of Prophecies,* Columbus elaborated upon his millennial vision and explained his role in it, supported by the prophecies of Daniel, Ezekiel, Isaiah, and, especially, John of Patmos. "And I saw a new heaven and a new earth: for the first heaven and the first earth were passed away; and there was no more sea," wrote John in the Book of Revelation (21:1). "God made me the messenger of the new heaven and the new earth of which he spoke in the Apocalypse of St. John after having spoken of it through the mouth of Isaiah," wrote Columbus, "and he showed me the spot where to find it."[26]

Columbus, master of the marine arts, thus identified his epoch-making technical achievement with the ultimate destiny of mankind. To his eyes, the discovery of the New World signaled the imminent End of the World, and hence the promised recovery of perfection. Identifying the Orinoco as one of the four rivers of the Garden of Eden, Columbus repeatedly insisted that he had indeed recovered the earthly paradise. "I am completely persuaded in my own mind," he wrote, "that the Terrestrial Paradise is the place I have said." And in the manner of a new Adam, he obsessively named all that he surveyed, confident in his expectation that mankind's original dominion might soon be restored.[27]

CHAPTER 3

VISIONS OF PARADISE

If millenarian expectations inspired the opening up of the New World, that opening further excited and confirmed such expectations, especially on the part of those Renaissance humanists and magi who, in the name and interest of religious revival, promoted the further advancement of science and the useful arts.

The new spiritual men of the fifteenth and sixteenth centuries, heirs of medieval millenarianism and precursors of the Reformation, sought in the study of nature and the recovery of ancient lore about the natural world the means of rekindling the true light of early Christianity. Thus the great humanist scholars Marsilio Ficino and Pico della Mirandola labored to unearth the lost secrets of hermetic natural philosophy and the occult arts, in the view of the Joachimite Augustinian abbot Egidio of Viterbo, as "messenger[s] of divine providence who had been sent to show that mystical theology everywhere concurred with our holy institutions and was their forerunner." Both men earnestly studied occult predictions and sought to square them with biblical prophecy. Pico was an admirer of the great Florentine prophet Savonarola, a disciple of Joachim of Fiore.[1]

The Renaissance alchemists and illuminati who followed in the wake of these humanist pioneers pursued their wondrous work in the same spirit. Cornelius Agrippa, for instance, drew inspiration from Joachimite commentary and identified Joachim as an example of one who gained prophetical knowledge from the occult meaning of numbers. "Because of the darkness caused by Adam's sin," wrote Agrippa, "the human mind cannot know the true nature of God by reason, but only by esoteric revelation." With regard to knowledge of the useful arts, Agrippa echoed the now conventional medieval monastic themes about Adam's divine endowment and the possibility of restoring mankind to its original and rightful dominion. "It was precisely this power over nature which Adam had lost by original sin, but which the purified soul, the magus, now could regain." "Once the soul has attained illumination," he argued, "it returns to something like the condition before the Fall of Adam, when the seal of God was upon it and all creatures feared and revered man."[2]

This medieval legacy also inspired the legendary alchemist Paracelsus, founder of the practical medical science of pharmacology. Paracelsus was immersed in the eschatological ethos of his times, and associated with "spiritualist" millenarian friends. Like them, he "foresaw the dawning of the Joachimite age of the Holy Spirit in which nothing would remain hidden and the arts and sciences would attain their greatest perfection." For him, the alchemist belonged to the spiritual vanguard, as "one who brought things to their perfection." "Human nature," Paracelsus wrote, "is different from all other animal nature. It is endowed with divine wisdom, endowed with divine arts. Therefore we are justly called gods and the children of the Supreme Being. For the light of nature is in us, and this light is God." "Each craft," he explained, "is twofold: on the one hand, there is the knowledge that we learn from men, on the other hand, the knowledge that we learn from the Holy Ghost." "Study without respite," he urged his fellows, "that the art may become perfect in us."[3]

Seen by many as a prophet himself, Paracelsus studied biblical prophecy and wrote an admiring treatise on a pseudo-Joachimite manuscript in which he dwelled upon such millenarian subjects as the sins of the Church and the expectation of the coming of the Antichrist and the angelic popes. He also wrote his own book of prophecy,

Prognosticato, which closes with the Edenic image of a man reclining at ease beneath a tree with the sun of divinity shining brightly upon him. "When the end of the world draws near," proclaimed Paracelsus, in rapturous expectation of millenarian redemption and restored perfection, "all things will be revealed. From the lowest to the highest, from the first to the last—what each thing is, and why it existed and passed away, from what causes, and what its meaning was. And everything that is in the world will be disclosed and come to light." "Then," he cautioned, "the true scholars and the vain chatterers will be recognized—those who wrote truthfully and those who traded in lies. . . . Blessed be those men whose reason will reveal itself."[4]

Paracelsus's apocalyptic vision was shared by his contemporary, the great Nuremberg artist Albrecht Dürer, who also shared his enthusiasm for the arts. Nuremberg was a celebrated center of mechanical arts, home to many masters of metalwork, from gunsmiths and armorers to the makers of scales, measuring instruments, and compasses. Dürer was himself born into a long line of goldsmiths; under his father's tutelage, he became an accomplished artisan, and later he littered his masterpieces (such as *Melancholia*) with tools of the trades. Throughout his life, he sought arts and secrets of nature and strived to elevate the social position of artisans and artists.

But, as with Paracelsus, religious expectation was Dürer's essence. "If we peer into the depths of Dürer's soul," wrote one biographer, "we find that the noblest and most essential element in his character was the religious urge. . . . The religious urge is the unifying element in Dürer's being, from which his genius developed." For Dürer, his workshop was his monastery, "the field on which the battle of his struggling and tormented soul was fought." Following Ficino, he believed that "art comes from divine inspiration," and he "considered his artistic activity as a calling to the service of God." Dürer was a fervent believer in both astrology and prophecy; his "apocalyptic mood" reflected the upsurge of popular millenarianism, then centered in Bohemia, as well as the early rumblings of the Reformation. (A Catholic all his life, Dürer was nevertheless hopeful of Church reform, and followed Luther's career with great sympathy.) His "first great work," which no one ever commissioned him to do, was the remarkably vivid series of woodcuts illustrating the Apocalypse of St. John. Completed in 1498, six years after the first messianic voyage of Christopher Columbus, it brought to life as never before the promise of mankind's redemption.[5]

The discovery of the New World induced an impatience with the Old. In vastly extending the range of the Renaissance imagination, it made Europe appear ever more despoiled, damned, and doomed, and prompted millenarian dreams of taking flight from this waning world in quest of new beginnings. In the New World, eschatological expectations of renewed perfection came into earthly focus.

After Columbus, paradise became more than just a vision; it became a place. Columbus identified the New World as the Garden of Eden. The Franciscan mystic Geronimo de Mendieta portrayed New Spain as the future site of the Kingdom of God. Here the worldly and the other-worldly, the present and the future, converged, giving rise to a new kind of apocalyptic vision of salvation that was as much the result of human ingenuity as faith: utopia. The utopias of Thomas More, Miguel de Cervantes, and Francis Bacon, for example, were all particular places, albeit difficult to locate—remote islands protected by endless sea. And the blessed inhabitants of these islands of perfection—utopia, Barataria, and New Atlantis—had made their paradise themselves, through their piety, their monastic discipline, their fraternal communalism, and their devotion to the useful arts.

The utopian "yearning to bring heaven down to earth," as John Phelan noted, was greatly stimulated by the Reformation. A religious revival of unprecedented proportions, the Reformation excited and legitimated millenarian hopes as never before, and made them more respectable. Only now were the heretofore condemned second-century millenarian writings of Irenaeus of Lyons recovered and included among his works. Martin Luther, who had studied Joachimite Franciscan prophecy, "revived the apocalypse as a pattern of history, an illumination of events past and . . . prophecy of things to come," while identifying the reformers as the chosen people confronting persecution but destined to triumph in the end; in the fourteenth century, John Wycliffe identified the papacy as Antichrist; and in the fifteenth, the Cambridge friar John Bale "placed Antichrist's identification with the papacy in a historical scheme influenced by Joachim and based on the Book of Revelation." For many in the sixteenth and seventeenth centuries, the rupture in the Church signaled the coming apocalypse, the prophesied end of the world and recovery of paradise. The reformers' emphasis upon the literal interpretation of Scripture, moreover, together with the development and spread of printing technology, made the prophetic books of the Bible, and hence apocalyptic speculation, more accessible. The writings of Joachim of Fiore were first printed in Venice early in the sixteenth century, coincident with Luther's break with the Church, and in this time of cataclysm, his apocalyptic vision gained new currency and wide circulation, among revolutionaries and elite reformers alike.[6]

In the midst of this apocalyptically charged milieu, utopian speculation about the kingdom to come took on an air of immediacy. And within this context, the medieval millenarian project of technological advance became more urgent. Even Thomas More's original utopia, based upon an essentially monastic vision of an austere, pious, and disciplined egalitarian community, reflected the already elevated conception of the useful arts as a medium of salvation; in utopia, every man had to practice a craft. For the utopians of the sixteenth and seventeenth centuries, the spiritual emphasis upon the useful arts and technical advance became central. "The gods have given man intelligence and hands, and have made him in their image, endowing him with a capacity superior to other animals," declared Giordano Bruno at the end of the sixteenth century. "This capacity consists not only in the power to work in accordance with nature and the usual course of things, but beyond that and outside her laws, to the end that by fashioning, or having the power to fashion, other natures, other courses, other orders by means of his intelligence, with that freedom without which his resemblance to the deity would not exist, he might in the end make himself god of the earth." "Providence has decreed," Bruno argued, anticipating Francis Bacon, "that man should be occupied in action by the hands and in contemplation by the intellect, but in such a way that he may not contemplate without action or work without contemplation. [And thus] through emulation of the actions of God and under the direction of spiritual impulse [men] sharpened their wits, invented industries and discovered art. And always, from day to day, by force of necessity, from the depths of the human mind rose new and wonderful inventions. By this means, separating themselves more and more from their animal natures by their busy and zealous employment, they climbed nearer the divine being."[7]

The Dominican friar Tommaso Campanella was, like Joachim of Fiore, a native of Calabria, and "his ardent expectation of a new world was founded on a Joachimite structure of history." At the turn of the seventeenth century, Campanella led an abortive rebellion in an effort "to push on the inevitable apotheosis of history" and establish his ideal city on earth. Facing his inquisitors, he explicitly identified himself as the embodiment of Joachim's third age. Campanella's utopian "City of the Sun" "enshrined the worship of science and technology as principles of social development and moral perfection." In this fraternal community, a Christian commonwealth whose origins may be traced to the similar imaginings of Raymond Lully, Francis Bacon, and Giordano Bruno, every citizen was required to master at least one mechanical art, and an unusual respect was accorded the accomplished craftsperson. The Solarian educational system, moreover, combined training in the mechanical arts with that of the liberal arts, which was "intended to give them the wisdom needed to understand, and to live in harmony with, God's creation."[8]

The utopian enthusiasm of the Continental reformers Johann Andreae and John Comenius also reflected the renewed millenarian mood—and thus the medieval millenarian expectations which it radically rekindled. As the historian P. M. Rattansi observed, "their social, religious, and educational reform was based on the conviction that the millennium was at hand, and would be marked by the recovery of the knowledge of creatures that Adam had possessed in his innocence, and of the Adamic language which had given him power over all things." In the view of this latest generation of determined dreamers, like that of their medieval forebears, the aim of science and the arts was the restoration of mankind's primal knowledge, shared with God at the beginning but lost in the Fall.[9]

Andreae had studied the prophecies of Joachim of Fiore as well as those of Paracelsus and other illuminati, and he ardently believed both that the millennium, and hence the prophesied restoration of perfection, was imminent, and that the advance of science and the arts was essential preparation for it. In Andreae's utopia, "Christian-opolis," the mechanical arts were to be assiduously practiced by its four hundred inhabitants. "All of these [crafts] are done not always because necessity demands it," he explained, echoing Erigena, "but . . . in order that the human soul might have some means by which it and the highest prerogative of the mind may unfold themselves through different sorts of machinery, or by which, rather, the little spark of divinity remaining in us may shine brightly in any material offered." "There is the greatest need that we return to ourselves as often as possible and shake off the dust of the earth," Andreae argued. The practice of the useful arts, among other activities, allowed men "to return to themselves." In his educational-reform efforts, the millenarian Moravian bishop Comenius promoted the teaching of the arts for the same exalted, essentially spiritual, ends.[10]

The Continental utopianism of Comenius and Andreae achieved its fullest and most influential expression in the millenarian manifestos of the mysterious Rosicrucian Brotherhood, which were probably written by Andreae. According to these bold apocalyptic proclamations, the Rosicrucians aimed at nothing less than "the reform of the whole of mankind," through the purification and reunification of Christianity and the cooperative advance of scientific and technological knowledge. The advent of the Rosicrucian revival was marked by the sudden appearance of this new spiritual order, surpassing in learning even the Jesuits, who likewise dedicated themselves to the study of science and the arts. The new brotherhood considered itself the latest incarnation of the Joachimite *viri spirituales,* the "new voice" of "a new arising sun" determined and destined to bring about a "third reformation of religion." And their bold and alarming manifestos (the "Confessio" and the "Fama Fraternitatis"), which were described by their author, in the terms of the Book of Revelation, as "our trumpet," were indeed to have a profound and lasting influence upon the modern European imagination.[11]

The manifestos stressed the advancement of useful knowledge, in a manner that reflected both the monastic and millenarian traditions. The learning of the Renaissance, according to the Rosicrucians, signaled the start of a new era of enlightenment, in anticipation of the millennium, which constituted, at the same time, a recovery of Adam's divine powers. The Rosicrucians viewed themselves as the embodiment and vanguard of this last great age of divine illumination. The "Fama" proclaimed that God "hath raised men, imbued with great wisdom, who might partly renew and reduce all arts (in this our age spotted and imperfect) to perfection; so that finally man might thereby understand his own nobleness and worth." The "Confessio" likewise declared that "God hath certainly and most assuredly concluded to send a grant to the world before her end, which presently shall ensue, such a truth, light, life and glory, as the first man Adam had, which he lost in Paradise, after which his successors were put and driven with him to misery. Wherefore there shall cease all servitude, falsehood, lies, and darkness, which by little and little . . . was crept into all arts, works, and governments of men, and have darkened the most part of them. . . . All the which, when it shall once be abolished and removed, and instead thereof a right and true rule instituted, then there will remain thanks unto them which have taken pains therein. But the work itself shall be attributed to the blessedness of our age."[12]

The Rosicrucian manifestos urged the learned people of Europe to respond to the fraternal invitation of the order and to cooperate with it in its providentially inspired undertaking. "The reader is adjured," advised the "Rosa Florescens," a later Rosicrucian manuscript, "to study with the Rosicrucian Brothers the Book of Nature, the Book of the World, and to return to the Paradise which Adam lost." The Rosicrucian appeal aroused "frenzied interest" throughout Europe and provoked in response a "torrent of literature," a "river of printed works." However, on the Continent, such seemingly revolutionary proclamations were met primarily with suspicion, fear, hostility, and repression. The urgent millenarian message found a more sympathetic following, on the other hand, in seventeenth-century England, where it was to have its most enduring impact.[13]

CHAPTER 4

PARADISE RESTORED

Though the ardent millenarianism aroused by the Reformation remained marginalized on the war-torn terrain of Counter-Reformation Europe, it gained respectability in Britain. In the eyes of those imbued with, or simply moved by, the prophetic spirit, England became in the seventeenth century what Spain had been in the fifteenth and sixteenth: the ark of the New Jerusalem. Reliable redoubt of Protestant reformers against the papal Antichrist, England became a safe haven for Continental exiles who helped to forge its messianic identity and mission. Equally important, here, as nowhere else in the West before or since, the scriptural expectation of an earthly redemption had come to suffuse an entire culture.[1]

The Bible was first translated into English by the doomed radical reformer Wycliffe in 1382, with little effect. In the early decades of the sixteenth century, candidates for the priesthood were still forbidden to translate or even publicly read the Bible without express episcopal authority, which was seldom given. In 1535, William Tyndale was forced to flee to the Continent to complete his English translation, only to be burned at the stake there for his efforts. Although Tyndale dedicated his translation to Henry VIII, the king nevertheless tried to block its importation into England, and otherwise restricted popular reading of the Bible.[2]

In 1539, the so-called Great Bible, based upon the translations of Tyndale and Coverdale, was finally authorized by Henry VIII, but only for reading in church. More than two decades later, Calvin's Geneva Bible, translated into English in 1560 by his son-in-law William Whittingham, became the first English household Bible, upon which most people soon came to rely for their scriptural instruction and inspiration. Most Protestants, especially Puritans, used the translated Geneva Bible until the publication of the King James Version in 1611. From the late sixteenth century on, then, the people of England were able to turn directly to scriptural authority, both for guidance in their everyday lives and for an understanding of their appointed role in the divine plan.[3]

The great English historian Trevelyan estimated that "the effect of the continual domestic study of the book upon the national character, imagination and intelligence . . . was greater than any literary movement in our annals, or any religious movement since the coming of Augustine [of Canterbury]." Indeed, in the seventeenth century, a literal reading of the Bible, in particular the prophetic books of the Old Testament and the Book of Revelation of the New Testament, became "central to all arts, sciences, and literature." Certainly, the English social historian Christopher Hill cautioned, "we must differentiate between the Biblical idiom in which men expressed themselves, and their actions which we should today describe in secular terms. But at the same time, we must avoid the opposite trap of supposing that 'religion' was used as a 'cloak' to cover 'real' secular motives. This may have been the case with a few indiviuals but for most men and women the Bible was their point of reference in all their thinking," their common resource, authority, and inspiration.[4]

In this scripturally charged context, soteriology and eschatology—preoccupation with salvation and speculation about the endtimes—fired the collective imagination. And herein the monastic and millenarian conceptions of redemption, which had ideologically ignited the advance of the arts, crystallized as never before: the monastic idea of transcendence as a recovery of mankind's divine likeness, a restoration of Adamic perfection, knowledge, and dominion, a return to Eden, and the identification of the arts as a vehicle of such transcendence; the millenarian idea of transcendence through history, the linking of the future with the past, the New Jerusalem with the lost Eden, and the identification of progress in the arts as the mark and medium of millennial advance, the fulfillment of divine prophecy.

A good deal of theological reflection of the period focused upon the Fall, in the firm belief that it could be reversed. Much attention was given to the person of Adam, in order to understand what he, and hence mankind, had once been (and might once again become). It was taken as a given that Adam was the be-all and end-all of creation, that, because of his image-likeness to God, he stood apart from and above the rest of the world. By divine design and authority, he enjoyed a superiority and dominion over all other creatures, and complete control over nature. "It is difficult nowadays to recapture the breathtakingly anthropocentric spirit in which Tudor and Stuart preachers interpreted the Biblical story," historian Keith Thomas observed. For the theologians of the early modern period, Eden was "a paradise prepared for man in which Adam had God-given dominion over all living things." This total dominion was forfeited in the Fall, but "despite the Fall . . . man's right to rule remained intact." Theologians argued, in the manner of Roger Bacon, that God had already granted to fallen man (such as Noah and Solomon) the means to recover his rightful reign. "Contemporary theology thus provided the moral underpinnings for that ascendancy of man over nature which had by the early modern period become the accepted goal of human endeavor."[5]

Central to this interest in recovering Adamic dominion over nature was earnest speculation about the extent of Adam's knowledge about nature, and in particular his knowledge of the useful arts, in the belief that "the fateful intellectual decline which had begun with the Fall of Adam might at last be reversed." "The extent of Adam's knowledge occupied considerable space in Biblical commentaries on Genesis," one student of the period observed. Many commentators maintained that Adam "must have been created with perfect knowledge," and they employed great "exigetical ingenuity" to demonstrate the range of Adam's artistic prowess, which extended from simple gardening to metallurgy. In a commentary from 1601, for example, Nicholas Gibbens declared that "all lawful and profitable Arts were known and practiced by Adam."[6]

The image of Adam as all-knowing and all-capable was an inspiration to reformers intent upon advancing science and the arts. As Charles Webster noted, "Writers were concerned to give a vivid impression of the great power sacrificed at the Fall, in order to galvanize their contemporaries into an effort to restore the primitive condition." Adam was thus viewed as the archetype artisan, putting his knowledge to useful, practical purpose, and hence as the source of inspiration for experimental philosophy. According to Hill, biblical texts, especially from Genesis, were typically quoted in support of "improvement" and the advancement of arts and crafts. "Puritan attitudes to technology and agriculture," Webster wrote, "were developed in the context of speculation about the primitive condition of man. In the Garden of Eden Adam willingly submitted to the discipline of work and his labor was pleasant. Because of his obedience, he was given complete control over his environment, until the Fall. Then he and his descendants were punished by being condemned to irksome toil. . . . But God had permitted man to turn the situation to his advantage through penitent labor." Moreover, "the practical arts were God's gift to his undeserving children," whereby they might "return from the suburbs to paradise." "Such sentiments," Webster observed, "reflect the emergence of a social ethic which placed considerable emphasis on unremitting toil and which accorded high esteem to the manual arts." As one commentator, Walter Blyth, confidently declared in the middle of the seventeenth century, "Scripture, reason, and experience showeth how we may be restored to Paradise on earth if we can but bring ingenuity into fashion."[7]

Such confidence was rooted in the millenarian enthusiasm which also marked the epoch, fueled anew by the upheavals of the time. Millenarians like John Napier, Thomas Brightman, John Henry Alsted, and John Comenius all interpreted the turbulent events of the Reformation as the end-times predicted in the prophetic books of Daniel and Revelation. The German Alsted effectively conveyed the apocalyptic spirit of the Continent to England, especially through the millenarian writings of his disciple Joseph Mede, the dean of English millenarianism, and of Mede's students, among them Samuel Hartlib, who became a close collaborator of Comenius; Henry More, mentor of Isaac Newton; and the poet John Milton.[8]

Like the Rosicrucians on the Continent, such men viewed the advancement of knowledge and the global extension of exploration and trade as sure signs that the millennium was at hand—in keeping with the often cited prophecy of Daniel (12:4) that "many shall run to and fro, and knowledge shall be increased." In addition, they read Revelation as the script of their historical moment. As one scholar noted, "the British Reformation spawned a great revival of historical interpretation of the Book of Revelation. . . . Protestants as a group believed that they were living near the end of the world, during the time prophesied in Daniel and Revelation, and made their fight against the Pope that of the righteous remnant against the Antichrist."[9]

In this apocalyptically charged context, millenarianism, as William Lamont reminds us, was not restricted to the "lunatic fringe." Unlike on the Continent, here it "meant not alienation from the spirit of the age but a total involvement with it." In England, "interpreting Revelation was a task for the most advanced minds," as well as the most exalted—King James himself wrote his own commentary on Revelation, and later an official translation of Mede's enormously influential *Key of the Revelation* was produced by a member of Parliament and published by order of a committee of the House of Commons, with a preface by the prolocutor of the Westminster Assembly of Divines.[10]

Here the most learned men of the age, as Richard Popkin observed, "took seriously the injunction in Daniel that, as the end approaches, knowledge and understanding will increase, the wise will understand, while the wicked will not. They also took seriously the need to prepare, through reform, for the glorious days ahead. Their efforts to gain and encourage scientific knowledge, to build a new educational system, to transform political society, were all part of their millenarian reading of events. They needed to understand, to construct a new theory of knowledge, a new metaphysics, for the new situation, the thousand-year reign of Christ on earth, which was to be followed by a new heaven and a new earth." As Popkin noted, "It is striking how all-pervasive the theme was, and how influential it was. . . . Efforts to accomplish this great end are part of the making of the modern world and of the making of the modern mind." "So at length when universal learning has once completed its cycle," wrote Milton, "the spirit of man, no longer confined within this dark prison-house, will reach out far and wide, till it fills the whole world and the space far beyond with the expansion of its divine greatness."[11]

This unprecedented millenarian milieu decisively and indelibly shaped the dynamic Western conception of technology. It encouraged a new lordly attitude toward nature, reflecting the anticipated restoration of Edenic dominion, and the associated notion, "which was to become common, that the study of the natural sciences will be carried on as an appropriate and important millennial activity." The monumental studies of the period by Charles Webster have made it abundantly clear that such millenarian preparation had a decidedly applied, utilitarian thrust, emphasizing the enhancement of technological prowess in agriculture, husbandry, mining, metallurgy, chemistry, mechanisms, and navigation. "The technological discoveries of the Renaissance, particularly those relating to gunpowder, printing and navigation," Webster wrote, "appeared to represent a movement towards the return of man's dominion over nature. . . . The Puritans genuinely thought that each step in the conquest of nature represented a move towards the millennial condition." As Milton insisted, in the course of millennial advance, nature would not merely become known to man but "would surrender to man as its appointed governor, and his rule would extend from command of the earth and seas to dominion over the stars."[12]

"With the models of Eden and the New Jerusalem in mind," as Webster observed, and in dedication to the active fulfillment of prophecy, the apocalyptically inspired reformers of the age "framed programmes for the development of applied science." Foremost among them was King James's Lord Chancellor Francis Bacon, whose "writings came to attain almost scriptural authority." Perhaps more than anyone else before or since, Bacon defined the Western project of modern technology, and his bold vision was "framed with reference to the millennial expectation of man's dominion over nature." For Bacon, the sustained development of the useful arts offered the greatest evidence, and the best means, of millenarian advance, because they alone were "continually growing and becoming more perfect."[13]

Bacon is typically revered as the greatest prophet of modern science, but, as Lewis Mumford rightly insisted, for Bacon that always meant "science as technology." Bacon viewed science not simply as a speculative enterprise but as one rooted in the practical arts and dedicated to utility and invention—"the relief of man's estate." Bacon recognized more clearly than his contemporaries the great achievements that had already been made by mechanical artisans in shipbuilding, navigation, ballistics, printing, and water engineering and he accordingly developed a utilitarian idea of scientific enterprise that drew much of its strength from artisanal practice. "Truth and utility here are the very same thing," wrote Bacon, meaning that the perfect knowledge acquired through science was best measured by its usefulness.[14]

Bacon viewed practical knowledge of the arts as the key to the advancement of knowledge in general, and used the mechanical arts as the model for the reform of natural philosophy. Like Paracelsus, Bruno, and the Rosicrucians, he insisted upon the elevation and elite appropriation of the arts. He aimed to establish, as he put it, "commerce between the mind of man and the nature of things," so that the practical arts might nourish and, in turn, be "nourished by natural philosophy." "Let no man look for much progress in the sciences," Bacon wrote in his *Novum Organum*, "unless natural philosophy be carried on and applied to particular arts, and particular arts be carried back again to natural philosophy."[15]

Bacon thus sought to close the gap between technology and philosophy, noting scornfully that among philosophers "it is esteemed a kind of dishonor unto learning to descend to inquiry or meditation upon matters mechanical." Toward this end, he insisted that philosophers must overcome their elite disdain for the useful arts, and learn to deal with "things themselves," "mean and even filthy things," in order better to appreciate their value and appropriate their fruits. In his defense of the worthiness of the useful arts, Bacon forcefully reasserted the tradition begun long before by Erigena, Hugh of St. Victor, and Roger Bacon, and sustained, most recently, by Paracelsus, Bruno, and the Rosicrucians. And as was the case with Erigena when he rewrote the script of Capella's marriage of Mercury to Philology, here too the union of the mechanical and the liberal arts was understood as bringing the former up to the exalted level of the latter, rendering technology not only worthy of elite attention but closer to God.[16]

For, if Bacon's effort was utilitarian in emphasis, it was transcendent in essence. If Bacon believed that the useful arts were essential for the advancement of knowledge, he also thought, like his forebears, that the advancement of knowledge was essential for salvation and the promised restoration of perfection, "the entrance into the kingdom of man, founded on the sciences," as he described it, "being not much other than the entrance into the kingdom of heaven."[17]

Bacon's transcendent goal, like that of his medieval precursors, entailed the recovery of mankind's original image-likeness to God. As biographer Paolo Rossi described it, Bacon's overriding aim "was to redeem man from original sin and reinstate him in his prelapsarian power over all created things." In historian Frances Yates's words, Bacon sought "a return to the state of Adam before the Fall, a state of pure and sinless contact with nature and knowledge of her powers," in short, "a progress back towards Adam."[18]

Bacon was explicit and insistent about the perfectionist purpose behind his advocacy of the useful arts. The title of his magnum opus, *The Great Instauration*, signifies reform as "a restoration," "a radical renovation," "a rehabilitation of past glory and primeval bliss." Indeed, an anticipatory fragment written two decades earlier, considered one of the most personally revealing of Bacon's writings, is subtitled "The Great Restoration of Man's Dominion Over the Universe." In this earlier statement, Bacon explained that he aimed "to stretch the deplorable narrow limits of man's dominion over the universe to their promised bounds." Two decades later, he likewise explained, in the preface to *The Great Instauration*, that he sought to show how the mind of man "might be restored to its perfect and original condition."[19]

"Man by the Fall fell at the same time from his state of innocence and from his dominion over creation," he explained in his *Novum Organum*, but "both of these losses . . . can even in this life be in some parts repaired, the former by religion and faith, the latter by arts and sciences." "It is not the pleasure of curiosity, nor the quiet of resolution, nor the raising of the spirit, nor victory of wit, nor faculty of speech, nor lucre of profession, nor ambition of honor or fame, nor enablement of business, that are the true ends of knowledge," Bacon insisted in *Valerius Terminus,* "but it is a restitution and reinvesting (in great part) of man to the sovereignty and power (for whensoever he shall be able to call creatures by their true names he shall again command them) which he had in his first state of creation." "We are agreed, my sons, that you are men," Bacon wrote in his "Refutation of Philosophies." "That means, as I think, that you are not animals on their hind legs, but mortal gods."[20]

Bacon's bold biblically inspired vision reflected the exaggerated anthropocentric assumptions of his seventeenth-century Protestant faith, the conviction that "human ascendancy was central to the Divine plan." "Man, if we look to final causes, may be regarded as the centre of the world insomuch that if man were taken away from the world, the rest would seem to be all astray, without aim or purpose," Bacon wrote. In the same scriptural spirit, he counseled humility in the pursuit of knowledge and power, lest mankind repeat the sin of Adam, but he defended his grandiose enterprise by insisting that "it was not that pure and unspotted natural knowledge whereby Adam gave names to things agreeable to their natures which caused the Fall, but an ambitious and authoritative desire of moral knowledge, to judge of good and evil, which makes men revolt from God." Like Roger Bacon before him, Francis Bacon maintained that the biblical accounts of Noah, Moses, and Solomon, as well as the history of the useful arts, offered sufficient evidence for the belief that the restoration of mankind's original powers was part of the divine plan. To-

ward the end of his life, in his utopian *The New Atlantis,* Bacon
glimpsed the fulfillment of this destiny, a time when God and man
would once again become co-workers in creation. For Bacon this
was not fantasy but foresight, a certain vision based upon scriptural
authority. He was firmly convinced, moreover, by millenarian prophecy, that this recovery of perfection was not just inevitable but
imminent.[21]

Like so many of his English contemporaries, Bacon believed that
the millennium was at hand. Inspired by prophecy, he viewed the advancement of knowledge in his own time as confirmation of this expectation, as well as a means of hastening and preparing for the
glorious days ahead. Throughout his career, Bacon proclaimed his
apocalyptic conviction. The advancement and spread of knowledge,
he declared in his early work *Valerius Terminus,* "by a special
prophecy, was appointed to this autumn of the world; for, to my understanding, it is not violent to the letter, and safe now after the
event, so to interpret that place in the prophecy of Daniel where
speaking of the latter times it is said, 'many shall pass to and fro, and
knowledge shall be increased,' as if the opening of the world by navigation and commerce and the further discovery of knowledge should
meet in one time or age." In his *Novum Organum,* he repeated the
message: "Nor should the prophecy of Daniel be forgotten, touching
the last stages of the world—'many shall go to and fro, and knowledge shall be increased'—clearly intimating that the thorough passage
of the world . . . and the advancement of the sciences, are destined by
fate, that is, by Divine Providence, to meet in the same age." Bacon
placed this oft-cited passage from Daniel, together with a drawing of
a sailing ship, symbol of the age of discovery, on the title page of his
Great Instauration, which he described as an "apocalypse or true vision of the footsteps of the Creator imprinted on his creatures."[22]

Bacon's advocacy of the useful arts in the interest of advancing human knowledge was aimed above all at the fulfillment of the
millenarian promise of restored perfection. Like his Continental precursors and contemporaries, particularly the Rosicrucians, Bacon believed that the development of science as technology was a means at
once of illumination and redemption. For the esteemed wise men of

Solomon's House in *The New Atlantis*, who embodied Bacon's ideal of a beneficent scientific regime, the emphasis was on the mechanical arts. Yet, as Frances Yates suggested, these invisible inhabitants of utopia at the same time had an almost "angelic" aspect and "would appear to have achieved the Great Instauration of learning and have therefore returned to the state of Adam in Paradise before the Fall."[23]

Largely through the enormous and enduring influence of Francis Bacon, the medieval identification of technology with transcendence now informed the emergent mentality of modernity. This transcendent impulse was especially pronounced during the Puritan Revolution, a period of both great millenarian promise and early capitalist enthusiasm for improvement and invention—fertile ground for Baconian reform. The Puritan Baconians were deeply involved in trade, overseas colonial projects, agriculture, ironworks, and other technological enterprises, and their optimism about technological transcendence matched their confidence in millenarian redemption.

At the center of this Baconian reform effort was the German émigré and Cambridge graduate Samuel Hartlib, whose social circle and "Office of Address" inspired and coordinated Puritan scientific, technological, and educational activities for decades. Because of his Continental background and associations, Hartlib also served as a channel for Continental utopian thought, especially through his translations of Alsted, Andreae, Campanella, and Comenius, and, perhaps most important, his friendship with Comenius. Also, through his collaboration with the educational theorists John Dury and Comenius, he became a major proponent of Baconian educational reform.[24]

As a merchant involved also in animal husbandry and farming, and through his Prussian family interests in trade and the dye industry, Hartlib was well acquainted with the practical and pecuniary callings of commerce, agriculture, and industry. As a result, his interest in and advocacy of science and the arts was decidedly utilitarian. The same was true of many of his associates and disciples, prominent among them Gabriel Plattes (author of the Baconian utopian tract *Macaria*, often attributed to Hartlib), an inventor who devoted himself to husbandry and mining; William Petty, who had firsthand knowledge of many trades, including textiles, smithing, carpentry, and coach-making, and was himself a physician and an accomplished inventor of technical instruments and agricultural machinery; and John Wilkins, who designed an improved plow and had a great interest in mechanical devices.[25]

Accordingly, Hartlib's proposed "Office of Address" focused upon "matters of ingenuity" and "the most profitable inventions," in its effort to identify, classify, transform, create, and, above all, appropriate knowledge. Like Bacon himself, Hartlib was a promoter of "the practical pursuit of nature"—"all pious and useful knowledge," "wisdom and inventions"—and stressed that "the principles and Arts ... might be indeed of some solid use, profit, or service to mankind." Likewise, the Baconian educational-reform efforts, following the teaching of Comenius, emphasized the practical application of knowledge to everyday life, and focused upon training in mechanics, animal husbandry, navigation, surveying, mineralogy, architecture, and metalworking. One of the central features of the "Office of Address" was a "college of Noble Mechanisms and Ingenius Artificers"; at Hartlib's invitation, Comenius himself, who had been greatly inspired by his reading of Bacon, visited England briefly before the outbreak of the Civil War, with the express purpose of making Bacon's Solomon's House a reality through the establishment of scientific schools.[26]

If the Baconians directed their advancement of knowledge and learning toward practical ends, however, they did so, like Bacon himself, in pursuit of a transcendent purpose. "Puritan attitudes to technology and agriculture were developed in the context of speculation about the primitive condition of man" and how to restore it, Charles Webster noted. "The final reward for such exertions might be man's return from the suburbs into paradise." In his letters to Hartlib, for example, the Baconian clergyman John Beale cited the biblical precedents of Moses and Solomon in support of such recovery. "And as Man is thus by light restored to the dominion over his own house," wrote Beale, "so he is restored to a dominion over all the beasts of the field, over the birds of the air, and over the fishes of the sea. Here you must add the discovery or dominion over all the Workes of God ... and of all the Elements to take such guise as Man by divine wisdom commands." In the eyes of Puritan reformers, this restoration of perfection was assured by both history and prophecy. Their goal was to "repeat the experience of Solomon (who regained dominion) and complete the prophecy of Daniel."[27]

Hartlib himself fervently believed that, because of the Fall, mankind was required to learn the arts anew, but that their earnest development would prepare men for a return to prelapsarian grace. As he explained in a petition to Parliament in 1649, the overriding purpose of his reform proposals was "the repairing and fitting [of] the decayed human nature to the good of society by a universal propagation of all arts and sciences in their reality and proper lustre." He was convinced, moreover, that in the millennium medicine would restore mankind to its original immortality. "I would have you understand my Prognostication of the true universal Medicine," Hartlib wrote in his *Chymical Address* of 1655, "which shall serve not only Men, but also all Flesh, namely that there grows in Paradise a Tree, which is, and is called the Tree of Life, which in the glorious and long expected coming of Jesus Christ our God and Saviour shall be made manifest, and then it shall be afforded to men, and the fruits of it shall be gathered, by which all men and all flesh shall be delivered from death, and that as truly solidly, and surely, as at the time of the Fall, by gathering the fruit of the forbidden tree, we together with all flesh fell into sin, death, and ill. And this glory and great joy hath God reserved for us, that live in these latter days, and hath kept his good Wine until now. . . . I do foretell all physicians, that then their Physic shall be worth nothing; for another Garden will be found, whence shall be had herbs, that shall preserve man not only from sickness, but from death itself."[28]

The Puritan program of practical, universal education was inspired by the same redemptive spirit, deriving largely from the Pietist educational philosopher and ardent millenarian Comenius. John Dury, the other major proponent of Puritan educational reform, wrote his own "Clavis Apocalytica, or A Prophetical Key" in the manner of Mede, as well as a treatise based upon the Book of Revelation entitled "Israel's Call to March Out of Babylon into Jerusalem," and spent most of his energies striving unsuccessfully to unite the various Protestant churches in anticipation of the millennium. The same inspiration moved John Milton, Hartlib's most illustrious associate in educational reform, whose treatise "On Education" was written at Hartlib's request and dedicated to him. For Milton too education was above all else a means of redemption, a way of recovering mankind's original God-likeness. "The end then of learning," he wrote, "is to repair the ruins of our first parents by regaining to know God aright, and out of that knowledge to love him, to imitate him, to be like him."[29]

CHAPTER 5

HEAVENLY VIRTUOSI

Though their utopian yearnings remained unfulfilled, the Puritan Baconians nevertheless laid the ideological groundwork for the so-called scientific revolution of the seventeenth century. Both their utilitarian outlook and their millenarian mentality gave formative shape to the milieu of modern science. Like Bacon, the founders of the new scientific academies also tended to view science as technology, as a philosophical enterprise inextricably bound up in both method and purpose with the useful arts. And they too were inspired in their work by an apocalyptic spirit which held out the promise for fallen man of a return to Edenic grace and restored dominion over nature. As their confidence in their bold enterprise increased, they raised their sights even higher. For perhaps the first time, they began to imagine moving beyond a mere recovery of what had been lost to something more and something new, beyond Adamic knowledge to the whole of divine knowledge, beyond a restoration of original creation to the making of a new creation.

The institutional and cultural divide which by the nineteenth century came to separate science from technology had not yet developed at the time the first permanent institutions of science were formed. The pioneers of the emergent scientific academies were utterly imbued with the Baconian spirit of usefulness. The first such academy, the Italian Accademia dei Lincei, which was formed in 1603 but shut down in 1630, was restarted in 1660 with the expressed aim "to improve the knowledge of natural things and all Useful Arts, Manufactures, and Mechanical practices, Engines, and Inventions by experiments." The Royal Society of London was founded that same year, with similar expectations.[1]

This orientation was already abundantly evident in the efforts of the various scientific circles which eventually coalesced to form the Royal Society, as well as in the views of their leaders who became the society's founders. The so-called Invisible College which formed around Robert Boyle pursued investigations in metallurgy, agriculture, and surveying largely for the profitable exploitation of Irish plantations. Inspired by these efforts, Boyle wrote his treatise on the "Usefulness of Natural Philosophy," in which he emphasized the methodological value for natural philosophers of "a real acquaintance with nature" through an involvement in practical pursuits, and, conversely, the ultimate usefulness of natural philosophy for the advancement of such practical concerns as navigation, husbandry, tanning, dyeing, brewing, metal-casting, pharmaceuticals, and warfare.[2]

The Oxford Club, which centered on John Wilkins, likewise emphasized practical investigations. Wilkins himself wrote a treatise on the importance of natural philosophy to husbandry, and was himself a mechanical inventor. The scientific circle associated with Gresham College also devoted themselves to practical ends, focusing primarily upon methods of navigation for merchants and seamen.

John Evelyn, the man who first proposed to Robert Boyle the formation of the Royal Society, later described its aim as simply "to improve practical and experimental knowledge," and this basic outlook was shared by the other original founders, including Wilkins; William Petty, an inventor in his own right and director of the Down Survey in Ireland; Christopher Wren, a prolific inventor and architect; and Henry Oldenburg, longtime society secretary, who was John Dury's son-in-law. There was also a strong connection between the scientific pioneers and early capitalist enterprise. Evelyn's family held the saltpeter monopoly; Boyle's father had interests in ironworks in Ireland; Petty was the son of a Romsey weaver and dyer with his own entrepreneurial interests; other early Royal Society members were involved in such industries as tobacco, distilling, and trade.[3]

The actual work of the Royal Society reflected these interests. Among the first permanent committees of the society were the mechanical, astronomical and optical, anatomical, chemical, and surgical sections, as well as a committee devoted to the history of trades. Early researches focused upon the practical problems of navigation (magnets and compass, maritime maps, hydrography, determination of longitude and latitude, times of tides, ship construction, hydrodynamics); mining (methods of ore extraction, water pumps and studies of atmospheric pressure, mine ventilation and air compression, metallurgy); military (iron-casting, gunpowder, trajectory and velocity of projectiles, recoil and range studies of arms, compression and expansion of gases, strength, durability, and elasticity of metals); textiles (wool manufacture, dyeing, silk-making, hat-making, watermills, windmills, and other "labor-saving" engines). "Thus the Baconian emphasis on the utilitarian applications of science was present from the beginning," Lewis Mumford insisted, "despite professions of detachment, neutrality, studious isolation, theoretic 'other-worldliness.'"[4]

But the professions of God-like detachment and neutrality, monklike isolation, and transcendent other-worldliness were there too, in the millenarian pursuit of perfection that underlay the scientific enterprise. As Margaret Jacob observed, the "millenarian impulse must be reckoned as one of the main motivations for the cultivation of scientific inquiry in seventeenth century England. . . . Almost every important seventeeth century English scientist or promoter of science from Robert Boyle to Isaac Newton believed in the approaching millennium." And at the heart of this millenarian expectation, and the ascetic dedication to learning it engendered, was the by now long-standing hope of recovering the Adamic knowledge lost with the Fall.[5]

As a young man in Geneva, Boyle experienced a severe storm which he identified with the end of the world described in the Book of

Revelation. As the storm raged, he pledged that if he was spared he would henceforth dedicate himself to a pious and chaste life, and Boyle adhered to that pledge. For Boyle, who is usually identified as the father of both experimental science and modern chemistry, empirical investigation was a form of spiritual experience, and knowing was at once a form of worship and an anticipation of millenarian resurrection. In his "Usefulness of Natural Philosophy," Boyle explicitly called for a renewal of Adamic knowledge, in anticipation of the millenarian recovery of perfection. "In the great renovation of the world, and the future state of things," Boyle wrote, "these corporeal creatures that will then be knowable shall probably be known best by those that have made the best use of their former knowledge. . . . And then the attainment to a high degree of knowledge, which here was so difficult, may, to the enlightened and enlarged mind, become as easy as it will be satisfactory." "To those virtuosi that are afraid to quit this world," Boyle added, "chiefly because they fear to lose the delightful philosophical knowledge they have of it, it may be represented, first, that it is likely that as our faculties will, in the future blessed state, be enlarged and heightened, so will our knowledge also be, of all things that will continue worth it, and can contribute to our happiness in that new state." Late in his life, in a treatise entitled "Some Physico-Theological Considerations About the Possibility of the Resurrection," he sought to explain resurrection in terms of the processes of chemical transmutation.[6]

The founding leaders of the Royal Society held similar views. John Wilkins saw the advancement of scientific knowledge as a means of mankind's recovery from the Fall and in his book on *The Beauty of Providence* expressed the millenarian theme that history would inevitably resolve itself into the "greatest serenity." Robert Hooke likewise declared that the specific purpose of the Royal Society was "to attempt the recovery of such allowable arts and inventions as are lost," and wrote his own continuation of Bacon's *New Atlantis*, in which he envisioned a future consolidation of religious, scientific, and political leadership in the hands of a Solomonic oligarchy "whose rule on earth corresponded to God's governance of the universe." Thomas Sprat believed that natural philosophical knowledge was an "excellent ground" to establish "man's redemption." Late in his life, John Evelyn wrote a treatise "Concerning the Millennium for the Countess of Clarendon" which revealed his own "fervent belief in the approaching apocalypse."[7]

Perhaps the fullest expression of this formative millenarian mentality of modern science was provided by Joseph Glanvill, a founding fellow and leading propagandist of the Royal Society, in his famous treatise in defense of the new science, The Vanity of Dogmatizing. In the manner of so many of his predecessors over the centuries, Glanvill began his book with a chapter entitled "What the Man Was" in which he described man's original Adamic powers lost in the Fall. "We are not now like the creature we were made," Glanvill reminded his readers, "and have not only lost our Maker's image, but our own." "All the powers and faculties of this copy of the Divinity, this medal of God, were as perfect as beauty and harmony in Idea," he wrote. "The senses, the Soul's windows, were without any spot or opacity. . . . Adam needed no Spectacles. The acuteness of his natural optics showed him most of the celestial magnificence and bravery without a Galileo's tube. . . . His naked eyes could reach near as much as the upper world, as we with all the advantages of arts. . . . His knowledge was completely built, upon the certain, extemporary notice of his comprehensive, unerring faculties. . . . Causes are hid in night and obscurity from us, which were all Sun to him. While man knew no sin, he was ignorant of nothing else." Such were the original endowments of mankind, which, Glanvill argued, might still be renewed by the "sons of Adam," along with their "ancient selves," through an appropriately pious and humble advancement of knowledge.[8]

"He that looks [for] perfection," wrote Glanvill, "must seek it above the Empyreum; it is reserved for Glory. . . . 'Tis no disparagement to Philosophy, that it cannot Deifie us [and] perfectly remake us after the Image of our Maker. And yet those raised contemplations of God and Nature, wherewith Philosophy doth acquaint us, enlarge and ennoble the spirit, and infinitely advance it above an ordinary level. . . . While we only converse with Earth, we are like it; that is, unlike our selves. But when engag'd in more refin'd and intellectual entertainments, we are somewhat more, then this narrow circumfer-ence of flesh speaks us." Glanville thus supposed that the great nat-ural philosophers, like Joachimite spiritual men of the third age, had already taken a step beyond mere mortals, and had come closer to their original God-likeness. "Upon the review of these great Sages, methinks, I could easily opinion, that *men* may differ from *men*, as much as *Angels* from unbodyed souls." He identified Descartes, Gassendi, Galileo, Brahe, and Harvey, among others, as "those gen-erous Vertuosos, who dwell in an higher Region than other Mortals," having partially recovered the "Image of *Omnipresence*. . . . These *Mercurial* souls, which were only lent to Earth to shew the world their folly in admiring it, possess delights, which as it were antedate Immortality, and (though at an humble distance) resemble the *joys* above."[9]

If the founders of modern science thus echoed the by now tradi-tional theme of a recovery of prelapsarian Adamic perfection—man-kind's image-likeness to God—through the advancement of knowl-edge and the arts, they also hinted, with increasingly more hubris than humility, at still loftier aims: the attainment of a truly divine un-derstanding of creation rather than its mere Adamic reflection, and the human extension, and hence fulfillment, of the divine project of creation itself. By "regaining to know God aright," as Milton mused, man might learn not only "to love him and to imitate him" but also "to be like him."[10]

Like Bacon, the scientists of the seventeenth century strongly be-lieved in the union of theory and practice. In their stress upon utility and craft-knowledge, as well as in their empirical and experimental emphasis, they insisted that the useful arts were not just the practical end of scientific knowledge but also the concrete means of gaining it, that if an understanding of nature could be put to practical use, so also putting nature to practical use was the surest route to such an understanding. In the course of his labors, the artisan gained not only a sound understanding of how his various mechanisms worked but also, perhaps, some sense of why they worked—an understanding of nature.

But such understanding remained fragmentary and indirect, a

mere by-product of art. In their quest for a more complete knowledge, the scientists sought a more direct connection with nature. They were inspired in this effort by a novel extension of what Amos Funkenstein described as their "constructive theory of knowledge." Increasingly, from Bacon on, knowing came to be viewed as a constructive process, the active result of making or doing something, rather than as a merely receptive process, the passive result of sensory impressions, reflection, or illumination. According to this view, true knowledge of something was the preserve of its maker, the artisan's sure knowledge of his artifact was the result of his having made it. This theory of knowing was now extended to knowledge about nature. A true knowledge of nature could only be gained through a recognition, first, of its having been made, of its existence as a creation, and, second, through an understanding of how it had been made, which was the privileged perspective of its Creator. "The Universe must be *known* by the *Art* whereby it was *made*," wrote Glanvill. Hence, in quest of such knowledge of nature—"the immutable workmanship of the omniscient Architect," as Boyle piously described it—the scientists raised their sights from Adam to his Father, from the image of God to the mind of God.[11]

Attempting to know the mind of God by scientifically deciphering the divine design behind nature, which now came to be viewed as a God-crafted mechanism, entailed a greater identification with God than did a mere recovery of Adam's divine image-likeness. As Amos Funkenstein explained, "the mechanical interpretation of nature could easily lead to the presumption that we know the making of the universe in the manner of the creator," and such a presumption "threatened to erode the wall between human and divine knowledge." Increasingly distancing themselves from both popular animistic lore and alchemical and hermetic philosophy, which assumed a divine presence in nature itself, the mechanistic scientists distinguished and divorced God from creation. They insisted upon a transcendent rather than an immanent deity, one who might yet occasionally tinker with his marvelous mechanism, which otherwise operated strictly according to preset laws, but only from outside it. Aspiring themselves to such a transcendent perspective, the scientists came to conceive "knowing as outside of nature," impersonal, detached, universal, abstract, and pure—and epitomized by mathematics. Scientific understanding, in short, became not merely Adamic but Godly.[12]

This new attitude was perhaps most evident in the work of Boyle and Newton. Although both men remained steeped in alchemical and hermetic philosophy and privately adhered to a spiritualistic appreciation (albeit ambivalent) of nature, they also embodied the ideal of the transcendent knower in quest of the transcendent God. Boyle learned Hebrew and other ancient languages in order to read God's words in their original expression. Likewise, as a natural scientist he delved directly into God's work—in the original, as it were—in an equally devout effort to come closer to his Creator. "It is the glory and prerogative of man," Boyle wrote, "that God was pleased to make him, not after the world's image, but his own." Thus, he urged, mankind must "look upon ourselves as belonging unto God." Boyle believed that this privileged relationship to God was especially embodied in the scientist, "born the priest of nature," whose "inquiry mediates between God and Creation." And he was convinced that, because of their great learning and devotion, the scientific virtuosi would ultimately, in the millennium, "have a far greater knowledge of God's wonderful universe than Adam himself could have had."[13]

Though Boyle no doubt identified himself, and was certainly viewed by his many acolytes, as the very model of the saintly virtuoso, the new transcendent ideal was more fully realized in the Godlike persona of Isaac Newton. Austere, ascetic, and aloof, Newton spent his entire life seeking some intimate understanding of his Creator. Like Boyle, Newton studied ancient languages better to understand the true meaning of Scripture. A fervent millenarian, he devoted a lifetime to the interpretation of prophecy, producing four separate commentaries on Daniel and Revelation. In Joachimite fashion, he believed that he "could prove, point by point, that everything foretold in the prophetic books had actually taken place, that the correspondence between prophecy and recorded history had been perfect." In a treatise on "The End of the World, Day of Judgement, and World to Come," he speculated about what the millennium and the

Kingdom of Heaven would be like, while he privately calculated the time of the second coming. Born on Christmas Day, he believed himself a messiah and a prophet (a status still accorded him by Seventh-Day Adventists) and wrote that the "Sons of the Resurrection" would have bodies like that of Christ, "with more than a touch of self-assurance that he would be among [them]."[14]

Whereas Boyle began his career emphasizing the usefulness of experimental natural philosophy only to argue later that "patient study was likely to enable man to gain a far larger share of his patrimony than aiming at immediate usefulness," Newton from the start "displayed sovereign indifference to the practical usages of science"; throughout his life his scientific efforts to discern the operating laws of nature were "directed almost exclusively to the knowledge of God." Newton's religious beliefs encouraged him "to search for divine efficacy in every aspect of the material order." For Newton, then, to uncover the hidden logic of the universe was to understand, and in that sense identify with, the mind of its Creator. Thus, as a scientist with divine pretensions, Newton had already begun his ascent. Indeed, according to his mentor Henry More, Newton "seemed to fancy himself soaring through the heavens . . . [which were] filled with a happy throng of saintly companions."[15]

To the deeply religious mind of modern science, beginning with Boyle and Newton (and also Galileo), the twin conceptions of "the divine transcendence of the creator-maker and the transcendence of man as knower reinforced each other." Henceforth nature was to be understood by the way it was made, which required of the scientist a God-like posture and perspective. But divine knowledge of creation was not all. Some aimed even higher, seeking not merely to know creation as it was made but also to make it themselves, actually to participate in creation and hence know it firsthand.[16]

In the sixteenth century, inventors and mechanics had increasingly invoked the image of God as craftsman and architect in order, by analogy, to lend prestige to their own activities: in their humble arts, they were imitating God and hence reflecting his glory. In the seventeenth century, the scientists began to carry this artisanal analogy between the works of man and God somewhat further, toward a real identity between them. Again, as Milton had written, they strove to know God not just in order to love and imitate him, but also "to be like him."[17]

The idea of man's participation in creation presupposed a belief that creation was not yet finished. This notion was rooted in the biblical belief in a "new creation," the expectation, based upon the promised advent of a second Adam, Christ, of man's redemption, the end of the fallen world, and the dawning of a new heaven and earth. Here God was not only creator, but also re-creator, refashioning his work to correct for its corruption by man. In the Joachimite millenarian scheme, man himself became through history a participant in his own redemption, and hence in the reconstruction of creation; through his mortal efforts, God completed his work. Thus, human actions, viewed as the expression of divine purpose through human agency, came to be seen as anticipations of the new creation, in fulfillment of the providential plan for the universe.[18]

For the millenarian reformers of the seventeenth century, the advancement of knowledge and myriad material improvements contributed to the completion of the first creation and constituted progress toward the new creation. Late in the century, the millenarian scientists likewise came to view their own useful designs and devices as extensions or augmentations of, and even improvements on, the original creation—a second nature, as it were—the human (but divinely directed) complement to creation. "And these great discoveries, which God in late ages hath made," John Beale, doyen of the Royal Society, wrote to Boyle, "may give us many grounds of good hope, that God makes haste to finish some great work in a more glorious display of so much of his lustre, as is fit for this world." In like spirit, the Restoration theologian John Edwards asked, "In Natural and Mechanical Philosophy and all Sorts of Mathematics, who sees not the vast improvements that these latter times have blessed us with. . . . Shall Divinity, which is the great art of arts, remain unimproved? . . . We see Divine knowledge and learning have been continually on the increase and yet we are sensible they are not come to the Full, whence therefore we reasonably conclude, that there are to be farther and greater augmentations in succeeding ages . . . before the conclusion of all things."[19]

Increasingly, in the inspired imagination of the time, man's contribution to creation loomed ever larger in the scheme of things. Despite their caveats about the necessity of humility, and despite their devout acknowledgment of divine purpose in their work, the scientists subtly but steadily began to assume the mantle of creator in their own right, as gods themselves. Francis Bacon, for example, had insisted that man's mission to remake the world was in reality but "the footsteps of the Creator imprinted on his creatures." "God forbid that we should give out a dream of our own imagination for a pattern of the world," he declared. Yet, at the end of his life, in his *New Atlantis*, he predicted that men would one day create new species and become as gods—"the undeclared ultimate goal" of modern science, as Lewis Mumford put it.[20]

Boyle also insisted that God had a hand in all human achievements, guiding men "by directing them to those happy, and pregnant, hints, which an ordinary skill and industry may so improve." Yet he also hinted himself at new God-like powers beyond those granted to Adam. "And sure it is a great honour, that the indulgent Creator vouchsafes to naturalists, that though he gives them not the power to produce one atom of matter, yet he allows them the power to introduce so many forms . . . and work such changes among the creatures, that if *Adam* were now alive, and should survey that great variety of man's productions, that is to be found in the shops of artificers, the laboratories of chymists, and other well furnished magazines of art, he would admire to see what a *new world*, as it were, or set of things has been added to the primitive creatures by the industry of his posterity." In this view, the Fall came to seem almost a blessing in disguise, in that through the exertions it necessitated fallen man had begun not merely to recover what had been lost but actually to supersede his original endowment—indeed, perhaps even to attain the divine powers that had been denied to Adam.[21]

CHAPTER 6

THE NEW ADAM

In Newton and his followers, as well as the guardians of Boyle's legacy, the transcendent scientific and technological spirit of the seventeenth century survived into the eighteenth. Millenarianism remained widespread in England both before and after the turn of the century, particularly among those Anglican churchmen who championed the new science, such as John Tillotson, Richard Bentley, William Whiston, Thomas Tenison, Samuel Clarke, and Thomas Burnet. They accepted the new science because it demonstrated an ordered, providentially guided pattern in nature that reinforced social order and stability—including the Church's authority, which was in their view a necessary precondition for millenarian advance.[1]

Tenison, archbishop of Canterbury, together with Boyle's disciple John Evelyn, served as trustees of the Boyle lectures, a forum on the new mechanical science established through Boyle's bequest to guard both science and religion against skepticism. Prominent Boyle lecturers included Whiston, who offered some of the earliest public expositions of Newton's system; Clarke, who defended the Newtonian system against Leibniz; and Bentley, master of Trinity College, Cambridge, who used the new science to defend the faith against deism.

Clergyman and mathematician William Whiston had been New-ton's assistant and, later, successor at Cambridge. Like his mentor, he sought to harmonize science and religion through a rational con-ception of creation; in his treatise on *A New Theory of the Earth,* he argued that the biblical stories of the creation, flood, and final confla-gration could be explained both historically and scientifically. He too diligently studied the writings of early Christians and, like Newton, became a devotee of Arianism. Less discreet than Newton about his unorthodox religious beliefs, he was dismissed from his academic position, and later became a Baptist. An avid millenarian, he orga-nized a society for the revival of early Christianity and wrote exten-sively on biblical prophecy.

Thomas Burnet was, like Evelyn, a millenarian disciple of Boyle. He believed that in the millennium humanity would be redeemed, and that men would then be reunited with God's mind, enabled to "think God's thoughts." Like Boyle and Newton, he vividly described the disembodied saintly existence that would follow the millennium. "And the great Natural character of it, is this in general, that it will be Paradisiacal. Free from all inconveniences, either of external Na-ture, or of our own Bodies."[2]

This same millenarian mentality continued to inspire scientific minds well into the age of enlightenment. James Burnett, Lord Mon-boddo, the Scottish philosopher and pioneer anthropologist whose study of the origins of language and society prefigured Darwin, de-clared a decade before the French Revolution that "the species is to end in not many generations. There will be a 'convulsion' of Nature, which is to produce a new Heaven and another Earth, to be inhabited by a new race of men, more righteous and pious than the former, and who are therefore called saints." He too described this saintly ex-istence, and also how best to prepare for it. "Our future happiness must be purely intellectual, produced by the contemplation of the wis-dom, the goodness, and the beauty of the works of God," Burnett ar-gued. "Now in order to enjoy this highest pleasure in a future life," he explained, "a man must be prepared for it in this life. And it is not sufficient that he is not vicious or wicked, but he must have cultivated his understanding by arts and sciences, and so have prepared his mind for the more perfect knowledge which he will have in a future state."[3]

The great eighteenth-century scientist Joseph Priestley viewed the French Revolution as the very millennial "convulsion of Nature" Burnett had predicted, an event which, he assured John Adams, was "opening a new era in the world and presenting a near view of the millennium." "My opinion," he explained to Adams, "is founded altogether upon revelation and the prophecies; I take it that the ten horns of the great Beast in Revelations mean the ten crowned heads of Europe, and that the execution of the king of France is the falling off of the first of these horns; and that the nine monarchies of Europe will fall one after another in the same way."[4]

A Dissenting clergyman who ultimately became, through Arianism, a founder of Unitarianism, Priestley is best known for his pioneering studies of electricity and, especially, his discovery of a number of gases, including oxygen. He became a fellow of the Royal Society in 1766 for his studies of electricity, supported by Benjamin Franklin, and was later made a member of the French Academy of Sciences. Like Newton and Boyle, Priestley was a lifelong millenarian. Early in his life he too studied Hebrew, Syriac, Chaldee, and Arabic in order better to understand the meaning of the Scriptures, and later also wrote his own commentaries on both Daniel and Revelation. He drew from such study an apocalyptic faith in the millennial advent of Christ, which was to be preceded by the return of the Jews to their homeland, the collapse of the Turkish Empire, and the fall of the Antichrist. "The great, though probably calamitous, events that are before us," he wrote in 1793, citing Daniel and Revelation, as well as Newton and Whiston, would ultimately bring about "a state of great knowledge, virtue and happiness" culminating in the Kingdom of God and the ascension of the saints, "who will live with Christ for a thousand years." Priestley refused to try to predict exactly when the millennium would occur, insisting, following Francis Bacon, that prophecy could only be confirmed as it was fulfilled in history, not before.[5]

Throughout his life, Priestley insisted upon the complementarity between his scientific work and his religious beliefs. He maintained that "man was bound by fixed laws of causation, but a good God had created the universe and was directing it toward a termination which would be the ultimate good of all." "The more I contemplate the great system, the more satisfaction I find in it," he wrote. "And the *structure* being so perfect, there cannot be a doubt but that the *end* and use of it, in promoting happiness, will correspond to it."[6]

True to his Baconian heritage, Priestley, like his predecessor Boyle, emphasized in all his work the practical application of science to the arts, for the purpose both of immediate utility and millennial preparation. As a student and professor in Dissenting academies, which were devoted to "education for practical life," Priestley always associated with and taught people destined for work in industry and commerce, who were denied entry to either the universities or the learned professions. He himself married the daughter of an ironmaster, and later became a member, together with industrial pioneers James Watt, Matthew Boulton, and Josiah Wedgwood, of the Birmingham Lunar Society, which was established to promote the application of science to industry and the crafts. Late in his life, living in voluntary exile in Pennsylvania, he cut back on many of his activities in order to devote his diminished energies to theology. He refused to abandon his chemical studies, however, because, as he explained, echoing Boyle, "I consider them as that study of the works of the great Creator, which shall resume with more advantage hereafter."[7]

Well into the age of enlightenment, and beyond, prominent promoters and practitioners of the arts and sciences continued to betray the transcendent visions and millenarian enthusiasms that had first inspired the Western enchantment with technological salvation. Michael Faraday, father of the modern understanding of electricity, who formulated the theories of electromagnetism and induction and invented the dynamo, was an elder and outspoken member of the Sandemanian sect of Christian fundamentalists who "lived by a highly literal interpretation of the Bible." For Faraday too science entailed both pious service to, and a devout effort at identification with, divinity.[8]

James Clerk Maxwell, the mathematician who gave Faraday's theories enduring and useful mathematical expression, was similarly inspired but less outspoken. A devout Christian, in private he studied the Bible and commentaries of the divines and wrote his own daily prayers. For one day in 1865 he wrote: "Almighty God, who hast created man in thine own image, and made him a living soul that he might seek after Thee and have dominion over Thy creatures, teach us to study the works of Thy hands that we may subdue the earth to our use, and strengthen our reason for Thy service, and so to receive the blessed Word, that we may believe on Him whom Thou hast sent to give us the knowledge of salvation."[9]

Charles Babbage, the mathematician, inventor, and pioneer of industrial automation, who is recognized as the father of the computer, maintained that mathematics and especially the "mechanical arts" supplied "some of the strongest arguments in favor of religion." "It is possible," he wrote, "that the advancement of man in the knowledge of the structure of the works of the Creator, might furnish continually increasing proofs of its authenticity; and that thus by the due employment of our faculties, we might not merely redeem revelation from the ravages of time, but give to it a degree of force strengthening with every accession to our knowledge. . . . By the exertion of the highest faculties with which we have been blessed, we may make a nearer approach to the knowledge of the will of our Creator." Babbage himself used the example of his Calculating Engine to demonstrate the probability, and hence the truth, of miracles, in particular the miracle of resurrection. In the manner of the seventeenth-century virtuosi, Babbage was inspired by a vision of immortality—"the after stage of our existence"—which was "founded on an instinctive belief that we are destined to be immortal by the Creator." When, "in a future state . . . with increased powers," he wrote, we could "apply our minds to the discovery of nature's laws, and to the invention of new methods by which our faculties might be aided in that research, pleasure the most unalloyed would await us at every stage of our progress. . . . Unclogged by the dull corporeal load of matter which tyrannizes even our most intellectual moments, and claims the ardent spirit to its unkindred clay, we should advance in the pursuit . . . [with] the irresistible energy resulting from the confidence of ultimate success." The other great pioneer of automation, Jacques Vaucanson, who devised the world's most ingenious automata, invented the prototype of the automatic loom, and served as the French inspector of manufactures, started his career as a Minim monk. No doubt reflecting his own preoccupation with regeneration, Vaucanson actually began work on "a completely artificial man," destined for both perfection and immortality.[10]

Perhaps more important historically, this same spirit now came to be embodied by another generation of spiritual men and institutionalized anew in yet another brotherhood of the sons of Adam, the Freemasons. The Freemasons became renowned for their strident defense of the dignity of the arts and their ardent promotion of useful knowledge—as well as their exclusive associations, core beliefs, and esoteric rituals. Following in the footsteps of monks, friars, illuminati, and virtuosi, the Freemasons carried the perfectionist project of the religion of technology into a more secular age, where they, in turn, passed it on to the new Adam of modernity, the engineer.

Modern Freemasonry evolved out of the guilds of medieval stonemasons, on the one hand, and the occult association of the Rosicrucians, on the other. From the first, Freemasonry derived a mythic identification with artisanry—the "craft"—and a dedication to the useful arts; from the second, it inherited its symbolic rituals and oaths, hermetic language and lore, a defining interest in the recovery of ancient knowledge and the advancement and diffusion of new knowledge, and "an all-pervasive religiosity." In the tradition of Andreae, Comenius, Hartlib, and Boyle, the Freemasons embraced the Baconian vision of "Solomon's House," a temple of divine knowledge dedicated at once to the relief of man's estate and the restoration of perfection. "They have built a Temple to wisdom," one Masonic orator proclaimed, "to renew there with all free and loving souls, the contract of the primitive fraternity."[11]

The early history of Masonry remains obscure and is clouded by many layers of Masonic mythology. Medieval lodges were established as resting and meeting places for itinerant stonemasons. This so-called operative Masonry—because it involved practicing masons—concerned itself with such traditional craft matters as wages and working conditions. In seventeenth-century England a new form of "speculative" Masonry emerged, which was aristocratic in origin and membership and concerned itself with loftier matters. The historical evolution from operative to speculative Masonry transformed a guild into a secret society and shifted its focus from the actual craft of building to the "moral and mystical interpretation of building," grounded upon worship of the "Great Architect." Because of the social standing of its members, speculative Freemasonry conferred upon the craft a greater degree of respectability than ever before. At the same time, however, actual practicing masons all but disappeared from Masonic membership. Speculative Freemasonry thus fully reflected the simultaneous ideological elevation and elite appropriation of the useful arts earlier envisioned by Francis Bacon.[12]

Exactly when speculative Freemasonry arose is not known for certain, but it probably had Rosicrucian roots, and its members were clearly instrumental in the founding of the Royal Society. The Freemason *Constitutions,* written in the 1720s, reflected Rosicrucian influence and closely resembled the work of Comenius, and the Masons' use of secret passwords was reminiscent of Rosicrucian practice. At the start of the documented history of speculative Freemasonry, in the 1720s, one out of every four English Freemasons was a fellow of the Royal Society.[13]

The generally recognized leader of speculative Freemasonry, John Theophilus Desaguliers, was himself a fellow of the Royal Society, a Newtonian natural philosopher, as well as an Anglican clergyman. An exiled Huguenot, Desaguliers studied at Oxford and became both an avid scientist and an accomplished inventor and engineer. He lectured on hydrostatics, optics, and mechanics, and gave some of the first public presentations of natural philosophy, for which he became renowned. He invented the planetarium, which was based on the Newtonian system; investigated the application of steam to various manufacturing processes (he was an early advocate of the Newcomen steam engine); experimented with electricity; translated Vaucanson's treatise on automata; and was involved as a civil engineer in the construction of bridges and fortifications. He was awarded the Royal Society's Copley Medal for the industrial application of science, and became the society's "official experimenter" and curator. Apparently he was held in high esteem by Isaac Newton, who was then president of the society.

Desaguliers was also a member of the Spalding Gentlemen's Society and lived and lectured at Bedford Coffee House, Covent Garden, both sites of early Freemason, and Royal Society, activity. The Spalding Society had been established for the promotion of industrial science by Royal Society fellow Maurice Johnson and the physician William Stukeley, who wrote millenarian tracts on "The Creation" and "Solomon's Temple," based upon the prophecy of Daniel. It was here that the millenarian William Whiston lectured on the industrial applications of Newtonian philosophy. Inspired by such associations, Desaguliers became a Mason and, by 1719, the third grand master of the English Grand Lodge. In this capacity, he initiated a thorough-going regeneration of Freemasonry by collecting old lore and commissioning the writing of the Freemason *Constitutions,* which has remained ever since the documentary cornerstone of Freemasonry.[14]

The opening sentences of the *Constitutions* echo the redemptive refrains of the religion of technology. "Adam, our first parent, created after the Image of God, the Great Architect of the Universe, must have had the Liberal Sciences, particularly Geometry, written on his Heart; for ever since the Fall we find the principles of it in the Hearts of his offspring. . . ." Following medieval tradition, the *Constitutions* traces the history of the ancient discovery, loss, and gradual recovery of the Adamic arts, and identifies Freemasonry as the modern medium of renewed perfection.[15]

> Hail Masonry! Thou Craft Divine!
> Glory of Earth, from Heaven revealed.
> Which dost with Jewels precious shine,
> From all but masons' eyes concealed.[16]

Thus does the "Fellow-Crafts Song," sung at the Grand Feast in all lodges, proclaim the sacred, privileged milieu of Masonry. Masonic mysticism and ceremony, devised in imitation of ancient, especially Egyptian, ritual, served at once to initiate the elect into its mysteries and to foster a sacred bond among its members in their "fraternal search for human perfectibility." "Through the sharing of common secrets and of a common language of signs, passwords, and hand-clasps," as one anthropologist described Masonic rituals, "through sharing the humilities of the ceremonials of initiation, through mutual aid, the frequent communion in worshipping and eating together, and the rules to settle disputes amicably between them, the members are transformed into a true brotherhood."[17]

Masons were not born but were "made" through rites of initiation. Initiates underwent a kind of symbolic regeneration or resurrection, whereby they overcame some of their mortal limitations. This belief (which perhaps accounts for some of the Masonic enthusiasm for Vaucanson's automata) was ritually enacted by having initiates experience death and rebirth by lying for a time in coffins and passing through forbidding labyrinths before being admitted into the illuminated company of Masonic membership. (This ritual is well illustrated in the Mozart Masonic opera *The Magic Flute;* Mozart was himself an enthusiastic Mason.) The experience of regeneration was not restricted to initiates but was an ongoing process. At least one lodge, in Strasbourg, held séances to search for "regeneration," and Masonic literature speaks consistently of the perfectibility of its members, an ascent which is reflected in a hierarchy of grades or degrees, twelve in all, from apprentice to sublime philosopher, each entailing its own wardrobe, jewelry, ceremony, mysteries, and measure of respect.[18]

Though Freemasons strictly avoided religious sectarianism and tended toward anticlericalism (they considered themselves the new priesthood), they were nevertheless in their own way devoutly religious. The fundamental requirement for membership was a monotheistic belief in and worship of the Great Architect of the Universe; a Bible always sat upon the Masonic altar, and distinctly Christian prayers opened and closed all meetings. (In the midst of the anti-Masonry movement in early-nineteenth-century America, Freemasons vehemently insisted that Masonry had always been "the handmaid of religion," devoted above all to divine worship. As one group of Masons declared, "Masonry requires a belief in the existence of an overruling Providence. Its forms and ceremonies are of a religious character. The Bible, in all Christian countries, is placed upon its altars, and in many of our prayers we invoke the name of Christ.") Masonic millenarianism, the elite pursuit of the "Masonic paradise," typically took the form of a secular utopianism. In the right context, however—such as late-eighteenth-century France—it could become distinctly apocalyptic.[19]

The Freemasons believed that their special knowledge was heaven-sent, but also that they were uniquely charged and morally bound to undertake its earthly development and dissemination—especially through the development and diffusion of the useful arts and sciences. In unleashing the redemptive powers of technology, the Freemasons were ahead of their time in practice yet ideologically bound to a long-established millenarian tradition. The Masonic mission, reflecting the Baconian spirit out of which it arose, was a decidedly practical enterprise turned to perfectionist ends. "Paradoxically," Margaret Jacob wrote, "a man rose to higher and more ornate and mystical status within a lodge because of practical virtue." Eighteenth-century Masonic diaries, she found, routinely associated "improvement" with "salvation." With a zeal inspired by such perfectionist passions, the Freemasons dedicated themselves to the advancement of the useful arts.[20]

Though the earliest Masonic missionaries, like Desaguliers, were Newtonians, they, like most natural philosophers of the day, eschewed the God-like detachment of their mentor in favor of an engaged utilitarianism. "In the eighteenth century," Margaret Jacob has suggested, "European Freemasons played a role in relation to scientific education analogous to that of progressive Calvinists in the seventeenth. In disproportionately large numbers, Freemasons promoted the new science by organizing lectures and philosophical societies for scientific devotees like themselves. In so doing, they exercised a role as progressive improvers, as the concrete promoters of the highest of Enlightenment ideals." Freemasonry was thus "the dynamic force behind the 'enyclopedias,' 'the diffusion of the light of knowledge,' and the promotion of the 'useful arts and crafts.' "[21]

The English, who were a generation ahead of their Continental brethren in the mechanical application of scientific knowledge, were the real pioneers in this regard, and the Freemasons were their vanguard. In 1755, William Shipley founded the Society for the Promotion of Arts, which later became the Royal Society of Arts and Crafts and the model for such efforts elsewhere in Europe, notably the

French Société d' Encouragement pour l'Industrie Nationale. The initial meeting of the society took place in the Masonic Bedford Coffee House, and the first president of the society were the Grand Master Earl of Morton; among the members were Desagulier's son, Benjamin Franklin, and other prominent Masons. Birmingham's Lunar Society, established in the following decade to encourage the industrial applications of science, included men of the same mold, among them the millenarian and Freemason Joseph Priestley. The encyclopedia movement typically associated with French *philosophes*, which had as its avowed aim the collection and diffusion of useful knowledge, also began in England with the Freemasons. The French *Grande Encyclopédie* was first conceived as a translation of the *Cyclopaedia, or General Dictionary of Arts and Sciences*, published in 1728 by the English Freemason Ephraim Chambers.

Freemasonry was introduced into France in the 1730s by Chevalier Ramsay, who became *orateur* of the Grand Loge de France in 1736. In his inaugural address, Ramsay declared that the order had as one of its overriding objects the diffusion of the useful arts. Ramsay's mention of Chambers's *Cyclopaedia* in this regard is viewed as the true starting point of the French *Grande Encyclopédie*. The French Masonic effort to promote the advancement of useful knowledge centered on the Masonic lodge La Loge des Neuf Soeurs, which, because of its distinguished international associations and commitment to educational reform, has been called the "UNESCO of the eighteenth century." According to its constitution La Loge was committed to the practical realization of the fundamental restorative aim of the religion of technology. "In making virtue its base," the lodge "has dedicated itself to fostering the arts and sciences. The aim of the lodge is to restore them to their place of dignity."[22]

In America, this same Masonic spirit of technological evangelism was evident in the educational-reform efforts of La Loge's grand master and lifelong Mason Benjamin Franklin. Franklin, who had been intimately associated with Masonry in both England and France, and was a member of perhaps the first Masonic lodge in America, was the foremost early promoter of the useful arts in America. His famous "Proposals Relating to the Education of Youth in

Pennsylvania," which led to the establishment of the Academy of Pennsylvania (later the University of Pennsylvania), was "the best-known early American argument for advanced training in the useful arts and sciences." Franklin's pioneering efforts were soon followed up, to considerable effect, by those of other Masons, including Grand Master DeWitt Clinton, educational reformer, father of the Erie Canal and the American internal-improvements movement, and the major force behind the American Society for the Promotion of the Useful Arts; Stephen Van Rensselaer, another champion of internal improvements and founder of the first civilian engineering school in America, Rensselaer Polytechnic Institute; and the prolific inventor and industrial entrepreneur Robert Fulton.[23]

Freemasons were also in the vanguard of early industrialization in Prussia, including eminent entrepreneurs and civil servants and military officers charged with technical or scientific matters. Among the entrepreneurs were Friedrich Dannenberger, Berlin's leading cotton manufacturer; Johann Hempel, owner of the city's largest chemical works; and a number of prominent Silesian industrialists. Masonic civil servants included Theodor von Schon, first head of the Business Department; Sigismund Hermbstadt, the chemical expert on the Technical Deputation; Ludwig Gerhard and Carl Karsten, the chiefs of the mining and metal divisions of the Mining Corps; and Christian Rother, chief of the Seehandlung, the state's merchant-banking empire. Military Masons included Johann Nepomuk Rust, chief of the Army Medical Corps, and the army's leading technologists, who were in charge of weapons testing, metallurgy, telegraphy, explosives, and army engineering.[24]

If the Freemasons were among the earliest advocates of industrialization, perhaps their most lasting and important, and heretofore unexamined, role was as midwives in the birth of the latest incarnation of spiritual men, the engineer. For engineering emerged as much out of Masonry as it did out of the military (indeed, the military itself was rife with Masonry). As the founding fathers of both the engineering profession and engineering education, the Freemasons passed on the legacy of the religion of technology to modernity's "New Man."

The modern profession of engineering, initially called "civil" engineering to distinguish it from a military function, first emerged, like Freemasonry—and from Freemasonry—in England. Among the earliest civil engineers were Grand Master Desaguliers himself, the leader of speculative Freemasonry, and John Grundy, one of the most important engineers of the first half of the eighteenth century, who was master of the Masonic lodge at Spalding. Thomas Telford, the practicing stonemason who became the "dominating figure" in the formative years of the engineering profession and "father" or "virtual founder of modern civil engineering," was also a Freemason (as was his lifelong collaborator, William Hazledine). Telford, who became the first president of the first professional engineering society, the Institution of Civil Engineers, founded his own Masonic lodge in Portsmouth while still in his twenties. "I take great delight in Free Masonry," he wrote in 1784, "and am about to have a Lodge Room fitted up at the George Inn, Portsmouth, after a plan of mine and under my direction."[25]

French civil engineering, like French Freemasonry, came into being a generation after its English counterpart, but it ultimately had a far greater influence on the emerging profession. Whereas the English Institution of Civil Engineers from its founding in 1818 played a major role in the training of engineers in England, the French became the real pioneers in professional engineering education and through it set the standard of engineering professionalism for the world. Here too the Freemasons were the central force.

The first professional engineering school, the Ecole des Ponts et Chaussées, was established by Jean-Rodolphe Perronet, the "father of engineering education." Perronet, the most renowned civil engineer of his time, was a member of the Uranie Lodge of Freemasons. The leading figure of the Ecole des Ponts et Chaussées after Perronet, Gaspard Riche de Prony, described as "the personification of the art of engineering," was also a Freemason, belonging to the L'Heuresse Réunion lodge of the Grand Chapter and the Chapitre Métropolitain. In Prony's view, the engineer was a new breed of man, the realization of the vision of the previous two centuries, during which "the science of the engineer began to experience the great development that prepared its current state of transcendence."[26]

The Ecole Polytechnique, which became the world's premier engineering school, was also the creation of Freemasons. The commission set up to formulate plans for its establishment was composed of four men—Antoine Fourcroy, Jean Hassenfratz, Claude Berthollet, and Gaspard Monge, all Freemasons. The mathematician Fourcroy and the chemist Berthollet were both members of La Loge des Neuf Soeurs; Hassenfratz, who came from the Ecole des Mines, belonged to the Le Bon Zèle lodge, and the mathematician Monge was the venerable first officer of the military lodge at Mézières, the Perfect Union of the Corps du Génie. Fourcroy drafted the commission's plan for establishing the new *école*, and Monge, the so-called father of Polytechnicians, became its guiding spirit.

Monge, the inventor of descriptive geometry, a fundamental contribution to modern engineering, was a professor at the famous military school, the Ecole du Corps Royal du Génie at Mézières. This institution, which had been founded in 1749, was equipped with laboratories for physics and chemistry and provided "the most advanced type of education offered in France and indeed in the whole of Europe." Here Monge developed his *répétiteurs* system of teaching descriptive geometry, which became the cornerstone of the educational experience at the Ecole Polytechnique. Here too Monge joined the Masonic lodge of which he became *orateur* and ascended to the Masonic degrees of Chevalier d'Orient and Rose-Croix, the latter a remnant of Freemasonry's Rosicrucian heritage.[27]

Besides becoming the legendary leader of the Ecole Polytechnique, Monge, together with his Masonic brethren, was involved in many other similarly inspired ventures. The duc de la Rochefoucauld, a member of La Loge, established a school for applied science in Liancourt, which was moved in 1803 to Compiègne and named the Ecole des Arts et Métiers (Arts and Trades). The course of study at the new school was formulated by a committee that included Monge and his intimate friend and Masonic brother Berthollet, both of whom were closely involved in the development of French manufactures. Around the same time, another group of industrially minded reformers, again including Monge and Berthollet, and also La Loge members Fourcroy and Jacques-Etienne Montgolfier, founded the Société d'Encouragement pour l'Industrie Nationale, modeled on the English and also Mason-inspired Society for the Promotion of Arts. Finally, the Ecoles Centrales, dedicated to the training of independent engineers on the British model, were created by legislation drafted, introduced, and administered by members of La Loge and their educational spirit derived from the pedagogical, and perhaps also Masonic, precepts of men like Gaspard Monge.[28]

Théodore Olivier, one of the founders of and later director of instruction at the Ecole Centrale des Arts and Manufactures, gave voice to this at once utilitarian and transcendent spirit in a celebration of *école-centrale* education, and tribute to Monge, entitled "Mémoires de Géométrie Descriptive, Théorique et Appliquée." "Man," declared Olivier, "forgetting that he is condemned to live on this earth, and dreaming only of the place where he will go after his terrestrial exile, places above everything his purely intellectual achievements. . . . The pure scientists thus forget . . . that work is the condition imposed on man." Man's divinely ordained task was to labor in imitation of the act of creation, to produce what Olivier described as "sublime and continually renewed modifications in the elements that form the terrestrial globe he inhabits." The students at both the Ecoles Centrales and the Ecole Polytechnique, meanwhile, expressed the Masonic legacy in a different way, through elaborate and intense initiation rituals.[29]

The French ideal of the engineer set the standard for the world. (Both Prussian and American—West Point—engineering education were expressly modeled on the Ecole Polytechnique.) Thus, through Freemasonry, the apostles of the religion of technology passed their practical project of redemption on to the engineers, the new spiritual men, who subsequently forged their own millenarian myths, exclusive associations, and rites of passage. If the Freemasons enshrined the Baconian gospel of harmonizing theory and practice—"As the Mechanical Arts gave occasion to the Learned to reduce the Elements of Geometry into Method, this Science, thus reduced, is the Foundation of those Arts," the Freemason *Constitutions* explained—the engineers embodied it. The incarnation of a thousand years of elite expectation, the engineers represented the renewal and elevation of the arts and personified the promise of technological transcendence.

The millenarian significance of the advent of the engineer was first announced by Henri Saint-Simon, the early socialist. Saint-Simon was closely associated with, and drew his disciples from, the Ecole Polytechnique, and had himself studied mathematics with Monge at the Ecole du Corps Royal du Génie. As social reformers, he and his followers became "evangelists for the engineer" and "apostles of the religion of industry," and ultimately forged a new religion, the New Christianity, on the basis of the Baconian vision of redemption from labor through science. But the true herald of the engineer was Saint-Simon's disenchanted disciple Auguste Comte.[30]

Comte, "a polytechnician through and through," had been a student at the Ecole Polytechnique—where he also studied mathematics with Monge—and throughout his life served as a tutor and admissions examiner for the school, in the vain hope of acquiring a professorship. It was Comte who provided "perhaps the most influential pronouncement on the question of the engineer's scientific identity." "An intermediate class is rising up," Comte wrote of the engineer in his "Third Essay," referring to Monge as the prime example, "whose particular destination is to organize the relations of theory and practice." Later he declared that "the establishment of the class of engineers in its proper characteristics is the more important, because this class will, without doubt, constitute the direct and necessary instrument of coalition between men of science and industrialists by which alone the new social order can commence."[31]

For Comte, who considered his "spiritual fathers" to be Bacon, Franklin, and Condorcet—and who was himself often described as the "Bacon of the nineteenth century"—the engineers constituted the vanguard of a positivist regime, "which has been rising since the time of Bacon." Comte believed that this new system would re-establish social order, in accordance with ineluctable natural laws, in the wake of the "crisis" of the French Revolution. It is quite remarkable, and testimony to the undiminished influence of the medieval ideological tradition, that, despite his vigorous iconoclasm and radical systemic critique of both theology and metaphysics, Comte reproduced almost in its entirety the millenarian mentality of the Middle Ages. "We [positivists] are the true successors of the great men of the Middle Ages," Comte proclaimed. Like Joachim of Fiore, he envisioned the movement of history deterministically as a succession of three inevitable stages—theological, metaphysical, and positivist—an understanding which came to him, as it had to Joachim, by way of revelation. Comte too described himself as a prophet and, as Frank Manuel observed, "noted on his manuscript the precise minute when the dynamics of the historical world were unfolded to him."[32]

For Comte, the advent of positivism represented the third, transitional stage—akin to Joachim's third age—which Comte described as the "transition towards the true and final doctrine" and the "total reorganization of society"—of which the engineers, as the new spiritual men, were the vanguard. If for Joachim the transitional third age entailed illumination by the Holy Spirit, for Comte it entailed a "restoration of religion" with the emergence of the "Religion of Humanity," "the final religion." "The positivist New Jerusalem is as definitely determined and measured as the Holy City of the Apocalypse," wrote one contemporary critic. "I am not exactly saying 'my kingdom is not of this world,'" Comte declared, "but the equivalent in terms of our epoch." Positivism, he maintained, "will afford the only possible, and the utmost possible, satisfaction to our natural aspiration after eternity."[33]

The overriding objective of Comte's positivist system was strikingly reminiscent of the Christian goal of a transcendent recovery of mankind's original divine image-likeness and dominion over nature. Science restores man to his place as "chief of the economy of nature . . . at the head of the living hierarchy," wrote Comte, with "pride of preeminence stirring within us, and above us the type of perfection below which we must remain but which will ever be inviting us upwards." Positivism was aimed at "awakening in all the noble desire of honorable incorporation with the supreme existence," and thereby attaining a "perfecting unity" with the Great Being, which would bring about mankind's "ultimate regeneration"—the "reconstruction" of "our whole nature," "the ultimate condition," the "definitive form of his existence," "the normal state."[34]

For the prophet of this positivist restoration of perfection, as for his millenarian forebears, the world's transformation was inevitable and imminent. All of history, Comte argued, revealed to him the "tendency toward regeneration," an inescapable movement toward the "kingdom of the Great Being," and "the normal state, the advent of which is shown by the whole past to be at hand." The "time is come," he proclaimed, for "the regeneration of the world by positivism," a transformation "as indispensable as it is inevitable. . . . No moral revolution ever existed at once more inevitable, more ripe, more urgent. . . ."[35]

A theist in spite of himself, Comte declared that the existence of the Great Being "is deeply stamped on all its creations, in morals, in the arts and sciences, in industry," and he insisted, as had previous like-minded prophets since Erigena, that all such manifestations of divinity were equally vital means of mankind's regeneration. No doubt because of his own training, associations, and intellectual proclivities, Comte was convinced that people like himself, science-minded engineering "savants occupied with the study of the sciences of observation are the only men whose capacity and intellectual culture fulfill the necessary conditions." These were the "priesthood of positivism."[36]

In preparation for their millennial role, like the monks, friars, magi, and virtuosi before them, the members of this new priesthood were urged to purify themselves through an abstention from worldly ambitions and a renunciation of the flesh. Comte's own asceticism, including his avoidance of tobacco and many foods and drinks as well as his postmarital celibacy, was legendary. "In his last years Comte's simplicity of living was such as might have won the approval of a medieval Franciscan," one biographer noted. Indeed, a positivist acolyte who visited Comte in 1851, six years before his death, observed that "he now reminded me of one of those medieval pictures which represent St. Francis wedded to poverty."[37]

This medievalism remained at the core of the positivist renovation of the world. Comte insisted that the "direct regenerative efforts made by the priesthood—efforts aiming at the preparation of the normal state and the reconstruction of the West"—were best achieved "by a worthy glorification of the past." If as a prophet Comte looked to the future and predicted, like Bacon and Milton before him, that through science and the arts mankind would ultimately become master even of biology and cosmology, he also maintained that medieval Catholicism provided the best model and inspiration for the new order. The remarkable degree of social order achieved by the Catholic Church, particularly in the Gregorian reforms of the eleventh century, was for Comte the great precedent to be emulated. Moreover, Comte counseled that until universal adherence to the religion of humanity was attained, "the mystical condensation of the medieval religion will serve as our daily guide in the study and improvement of our nature." Accordingly, he formulated a new catechism, new rituals of worship which bore striking resemblance to their medieval originals. And he earnestly advised his disciples to follow his example by reading daily from Thomas à Kempis's *Imitation of Christ*.[38]

Comte's technology-inspired millenarianism, albeit without its overtly religious excesses, was shared by nineteenth-century socialists, who also grounded their philosophical systems on an explicit rejection of religion. Like the Freemasons of the Enlightenment, the socialists carried the religion of technology into a more secular age; unlike the Freemasons, they rendered it a popular as well as an elite obsession. Robert Owen declared that "all religions of the world are founded in error," and that all religious people are deluded by "fables and doctrines" not of their own making. He also well recognized the human tragedy that typically attended the capitalist introduction of machinery. Yet he earnestly believed that a more humane use of machinery would prove liberatory, given an accompanying transformation in social relations and cultural habits, and accordingly "offered a millennial vision of the transformation of the machine." Invariably the Owenites came to view technology in terms of "promise and prospect" and as a determinant rather than a result of social transformation. The steam engine, they came fervently to believe, would "further advance that grand and growing cooperation [in the] far time of the Millennium." Power machinery became in their eyes "a god of the state of bliss," something to be worshipped.[39]

At length, casting away his guise of terror, this much cursed power revealed itself in its true form and looks to men. What graciousness was in its aspect, what benevolence, what music flowed from its lips; science was heard and the savage hearts of men were melted; the scabs fell from their eyes, a new life thrilled through their veins, their apprehensions were ennobled, and as science spoke, the multitude knelt in love and obedience.[40]

Later Karl Marx would level perhaps the most profound intellectual assault upon both religion—the "opiate" of the masses—and the capitalist use of machinery to degrade and enslave human labor. Yet, at the same time, he identified the technical development of the means of production as the underlying historical substrate of deliverance, laying the material basis not only for capitalist accumulation but also for the social revolution that would signal the end of class society and thus the transcendence of history. For Marx as for Owen, machines did not by themselves change society, only people did, but machines did promise (if only they were put in the right hands) an Edenic respite from labor. On that basis, Marxism evolved into "a near chiliastic hymn to a technological apocalypse" and became the most influential Western prophetic system since that of Joachim of Fiore.[41]

CHAPTER 7

THE NEW EDEN

If echoes of the religion of technology continued to resound in the clarion calls of European socialism, the true center of this enduring faith had by now shifted farther west, to the promised land of the New World. In America as nowhere else before or since, the useful arts became wedded to Adamic myths and millenarian dreams. To be sure, the Masonic influence played a part, evidenced not only by the early efforts of men like Benjamin Franklin, DeWitt Clinton, and Stephen Van Rensselaer but also by the remarkable, and heretofore unremarked, fact that Masons have been among the most prominent pioneers of every American transportation revolution: canals (Clinton and Van Rensselaer); steamboats (Fulton); railroads (George Pullman, Edward Harriman, James J. Hill); the automobile (Henry Ford); the airplane (Charles Lindbergh); and spaceflight (at least half a dozen astronauts, including John Glenn, the first man to orbit the earth, and Edwin Aldrin, lunar-module pilot for the first landing on the moon). And a technology-inspired millenarian socialism also contributed to the American vision, through the efforts of first Owenite and later Marxist émigrés, and, most notably, in the homegrown utopian writings of Edward Bellamy. But above all it was a genuinely and fervently religious spirit which fueled the American fancy.[1]

If Columbus identified the New World as the terrestrial paradise and the Franciscans viewed their missionary efforts there as the quickening of millenarian momentum, it was successive generations of millenarian Protestants who gave America its defining myth, rooted in the providential promise of new beginnings. "The American myth saw life and history as just beginning," R. W. B. Lewis suggested. "It described the world as starting up again under fresh initiative, in a divinely granted second chance for the human race." The hero of the myth was "a new Adam," "an individual emancipated from history [and] easily identified with Adam before the Fall." Here "progress toward perfection" was at the same time the recovery of "primitive Adamic perfection." The American was "the eternal Adam," who would create "an earthly millennium of perfect harmony in the New World Eden." "I, chanter of Adamic songs, through the new garden the West," waxed Walt Whitman. "Divine am I, inside and out, and I make holy whatever I touch."[2]

Edward Johnson's 1628 solicitation for volunteers to colonize New England fully reflected the earnest expectations of his time. America, he wrote, would be the place where "the amalgamation of the City of the World into the City of God" would take place. "For all your full satisfaction, know this is the place where the Lord will create a new Heaven, and a new Earth in new Churches, and a new Commonwealth together." In the same spirit, John White viewed this blessed land as "a bulwark . . . against the Kingdom of Antichrist," and Cotton Mather's reflections upon it "brought into his Mind the New Heaven, and the New Earth, wherein dwells Righteousness."[3]

A century later this myth was reaffirmed in religious revival, during the First Great Awakening. "The Millennium is begun," declared Boston minister John Moorhead. Likewise, Jonathan Edwards proclaimed with confidence in 1739 that "this new world is probably now discovered, that the new and most glorious state of God's church on earth might commence there; that God might in it begin a new world in a spiritual respect, when he creates the new heavens and the new earth." For Edwards, the revival signaled "the dawning, or at least a prelude, of that glorious work of God, so often foretold in Scripture, which in the progress and issue of it, shall renew the world of mankind." "Many things . . . make it probable that this work will begin in America," he added.[4]

And again, a century after that, during the even more intense Second Great Awakening, millennial expectation was renewed and reaffirmed by "militant Protestant Christianity." For the mass of the American democracy, wrote Perry Miller, "the decades after 1800 were a continuing, even though intermittent revival." "Rank on rank they advanced with flying banners," one contemporary described it, "the revivalists leading the way, the missionary societies, the Bible societies, the Sabbath reformers, the religious education and sabbath school societies, and the tract societies. Combined in the same great army, and under the same staff were the anti-slavery societies, the peace societies, the Seaman's Friend Society, the temperance societies, the physiological reform and moral reform societies. Closely allied were the educational reformers whose task it was to train a generation for utopia. In the heavens they saw the reflection of the glorious dawn, which was just beyond the horizon. . . . In America all things were being made new. In America where all was progress, development, movement and hope, in America the Millennium seemed about to begin. . . ." Or it had already begun.[5]

This, then, was the ideological context of technological development in America, where scientific and industrial revolutions followed in the wake of religious revival. The premillennialists earnestly anticipated and piously prepared for Christ's imminent return and the start of the millennium. The postmillennialists, believing that Christ would return only at the close of the millennium, which had already begun, righteously set about constructing his earthly kingdom. For both, the arts and sciences were means to millenarian ends: the making of the second creation. Here "second creation" meant that it was made by man, albeit divinely inspired and ordained, rather than by God directly; that its result was artifice, a secondary elaboration upon and extension of the first creation, nature; and that it reflected the arrival or imminent advent of the millennium, which marked a new genesis, a restoration of perfection, a new creation. In this context, the advance of the arts was at once the work of man and God, the useful human development of the earth and the reign of heaven on earth. As

Miller suggested, "it was not only in the Revival that a doctrine of 'perfectionism' emerged; the revivalistic mentality was sibling to the technological."[6]

Among the earliest Americans to make the association was Jonathan Edwards. " 'Tis probable that the world shall be more like Heaven in the millennium in this respect: that contemplation and spiritual employments, and those things that more directly concern the mind and religion, will be more the saint's ordinary business than now . . . [because] there will be so many contrivances and inventions to facilitate their secular business . . . , [including] contrivances for assisting one another through the whole earth by more expedite, easy, and safe communication between distant regions." This theme was repeated by later millennialists, including Joseph Bellamy (great-great-grandfather of Edward Bellamy) and Samuel Hopkins, who saw the millennium as a period of relative ease, a time, as Hopkins put it, of "outward conveniences and temporal enjoyments." "There will also doubtless be great improvements and advances made in all those mechanic arts, by which the earth will be subdued and cultivated, and all the necessary and convenient articles of life, such as all utensils, clothing, buildings, etc., will be formed and made in a better manner, and with much less labour, than they are now," Hopkins wrote in 1793, in his *Treatise on the Millennium*. In short, the millennium would usher in "a fullness and plenty of all the conveniences of life . . . [more] than ever before, and with much less labour and toil," a mechanized return to Eden.[7]

The musings of such millenarians were elaborated in great detail a half-century later by the German émigré, inventor, and civil engineer John Adolphus Etzler. Influenced by Hegelian philosophy (like his fellow émigré and friend John Roebling, chief engineer of the Brooklyn Bridge) and Owenite socialism, as well as by American evangelism, Etzler wrote in 1833, at the height of the Second Great Awakening, what was probably the first American technological utopia, a decidedly practical guide with the distinctly millenarian title *The Paradise Within the Reach of All Men, Without Labor, by Powers of Nature and Machinery*.

Etzler's Eden was based upon mankind's ability to reason, by the exercise of which he might regain his original bliss. "If man ever forfeited the paradise by his sin, as we are told," wrote Etzler, "it must have been the sin of neglecting the most precious gift of his maker, that reasoning faculty, that only gives him the dominion over the brutes, and may give him also the dominion over the inanimate creation, and make thereby of the earth a paradise. Man needs not to eat his bread in the sweat of his brow...." Technologically, the paradise was based upon the useful harnessing of the powers of nature—the wind and waves, the tides, and the heat of the sun—the technical means for which he described in great detail. Etzler was well aware, however, of Adam's sin of hubris. Though he often rhetorically referred to the God-like powers humans would gain thereby, he was careful to distinguish the arts of the second creation from those of the first. "Powers must pre-exist, they cannot be invented," he insisted, dismissing as futile folly the pursuit of perpetual motion. Humans cannot create these divinely given powers, they can only tap them. But in so doing, they can "cause a regeneration of mankind," and bring into being a "paradise" of peace and plenty, "a general state of sincerity, innocence, and true intelligence.... Mankind may thus live in and enjoy a new world, far superior to the present, and raise themselves far higher in the scheme of being." "It would seem from this," noted Henry David Thoreau in a review of the book, "that there is a transcendentalism in mechanics."[8]

According to one biographer, Etzler "sought the end of history in the accomplishment of his paradise on earth" and undertook a "messianic journey ... in search of the right conditions under which ... to re-establish the Paradise that Adam lost for mankind." If he preached, he also practiced, devoting his life to a succession of experimental ventures, at once technological and communitarian. "We are on the eve of the most eventful period of mankind," he declared, "a universal paradise of peace, abundance, happiness, and intelligence," "a new order of things," a "new world" in which men "may get a fore-taste of heaven" and be "so much the better prepared for another paradise hereafter."[9]

Eccentric in his enthusiasm, Etzler nevertheless well expressed the ethos of his era. A decade earlier, amid some of the most intense revivals of the Second Great Awakening, Amos Eaton, another American technological enthusiast, "wandered through the New England states and New York like a religious evangelist," preaching the Baconian gospel of the usefulness of scientific knowledge. A "devout Christian" who combined "the study of the Word and of the Works of the Creator," Eaton was more successful than Etzler. Under the patronage of the Masonic patroon Stephen Van Rensselaer, Eaton was able to put his preaching into practice as the guiding spirit behind the nation's first civilian engineering school, Rensselaer Polytechnic Institute.[10]

And only a few years earlier, Harvard professor Jacob Bigelow, whom Miller described as the "true prophet" of utilitarian science in America, published a series of lectures on the integration of science and the useful arts in which he introduced a new word: "technology." A thousand years after Erigena first coined the generic term "mechanic arts" to signify the arts and crafts in general, Bigelow, borrowing from German Professor Johann Beekmann's earlier use of the term in his encyclopedic history of inventions, gave a generic name to the "arts of science." As Erigena had identified the arts as a true reflection of mankind's image-likeness to God and a means of recovering Adamic dominion, so Bigelow revered the new body of learned men, "who heaven's own image wear," for reconciling "faith and truth," and he recounted the historic "restoration of the arts," which served "to extend the dominion of mankind over nature." Paying homage to "the mighty mind of Bacon," Bigelow boasted that "we have acquired a dominion over the physical and moral world, which nothing but the aid of philosophy could have enabled us to establish." "Next to the influence of Christianity on our moral nature," he told an audience at the opening of the Massachusetts Institute of Technology—which at his suggestion had adopted the new term for its name—"[technology] has had a leading sway in promoting the progress and happiness of our race."[11]

The appreciation of men like Etzler, Eaton, and Bigelow for what Leo Marx called the "technological sublime" was widely shared by their contemporaries. "These are only the precursors of other still more sublime accomplishments reserved for human genius," one American wrote of the uses of the steam engine—"the dawnings of that perfection which futurity will unfold." (George Wallis, a member of the British committee sent to investigate the American system of manufactures, noted wryly that for the Americans "only one obstacle of any importance stands in the way of constant advance towards greater perfection, and that is the conviction that perfection is already attained.") Machinery and transcendentalism "agree well," wrote Ralph Waldo Emerson."[12]

The advent of the telegraph, for example, "enter[ed] American discussions not as a mundane fact but as divinely inspired for the purposes of spreading the Christian message farther and faster, eclipsing time and transcending space, saving the heathen, bringing closer and making more probable the day of salvation." The very first message conveyed over this new invention was from the Bible: "What hath God wrought!" (a scriptural selection provided by the daughter of the U.S. commissioner of patents). The telegraph's inventor, Samuel F. B. Morse (whose staunchly evangelical father, the geographer Jedidiah Morse, had been among the founders of both the American Bible Society and the New England Tract Society), was a generous benefactor of "churches, theological seminaries, Bible societies, and mission societies." "The nearer I approach to the end of my pilgrimage," Morse reflected late in life, "the clearer is the evidence of the divine origin of the Bible, the grandeur and sublimity of God's remedy for fallen man are more appreciated, and the future is illuminated with hope and joy."[13]

We are on the "border of a spiritual harvest because thought now travels by steam and magnetic wires," exulted the preacher Gardner Spring. This same sentiment was exuberantly expressed in 1856 in a poem addressed to "Professor Morse":

> A good and generous spirit ruled the hour;
> Old jealousies were drowned in brotherhood;
> Philanthropy rejoiced that Skill and Power,
> Servants to Science, compass all men's good;
> And over all Religion's banner stood.[14]

The inspired incantations of the technological utopians thus resonated with, as surely as they reflected, widespread popular imag-

inings. And these were themselves but a dim reflection of the extravagant claims of the engineers themselves, the self-anointed priesthood
of the new era. The engineering mission, declared mechanical engineer George Babcock, is to bring about the time "when every force in
nature and every created thing shall be subject to the control of
man." "The civil engineer is the priest of material development," exclaimed civil engineer George S. Morison; "He is the priest of the new
epoch."[15]

Toward the end of his life, Morison, a prominent railroad engineer who became the leading bridge-builder of his day and a president of the American Society of Civil Engineers, wrote *The New
Epoch: As Developed by the Manufacture of Power,* a book reminiscent of Etzler's earlier treatise. As Etzler had based his paradise on
"the powers of Nature and machinery," so Morison based his on the
"manufacture of power." Also like Etzler, Morison was careful to
distinguish the man-made manufacture of power from the divine creation of its sources. "Creation, whether of substance or force, is not
given to man," wrote Morison; on the other hand, "manufacture is
not creation, but to change inert matter from one form to another in
such a way as to generate power is to manufacture power, and this
we can do." The new epoch was built upon such capability, an era
that the civil engineer "is bringing into existence by the manufacture
of power."[16]

Morison described this new epoch with the standard hyperbole.
"No changes have ever equalled those through which the world is
passing now," he wrote; "the new epoch differs from all preceding
epochs" and will create an entirely "new civilization." This epoch will
see the final and "inevitable" destruction of "savagery," "barbarism,"
"ignorance," and "superstition." And in its wake, "mankind must
settle down to a long period of rest," marked by "contentment,"
"comfort," and "happiness." Moreover, "it will not be the condition
of a town nor of a nation but of the whole earth, with nothing to
change it unless communication should be opened with another
planet."[17]

The engineer and engineering educator Robert Thurston seconded such sentiments. The son of one of the first American builders

of steam engines, Thurston became an expert on steam power (and steam-boiler accidents) and the leader of the new profession of mechanical engineering. He was a founder and first president of the American Society of Mechanical Engineers (ASME), president of the Stevens Institute of Technology, and founder and first head of the mechanical-engineering school at Cornell University. He also established the first mechanical laboratory for research in engineering, thereby institutionalizing the Baconian approach to the useful arts.

Reflecting upon the engineering enterprise at the turn of the twentieth century, Thurston wrote a remarkable series of articles, based upon an address given to the Pennsylvania chapter of Sigma Xi, the engineering honor society. "The truths of science and the truths of religion can never conflict," Thurston declared, no more than there can be "ground for conflict between those who seek to promote pure science and those who as earnestly and honestly endeavor to advance applied science." Indeed, he insisted, science and engineering afford "an increasing appreciation of, and familiarity with God's ways." Bringing full-circle the thousand-year history of the religion of technology, Thurston likened the scientific and engineering endeavor to "revelation and prophecy," which he suggested had now become "the fruits of science." "The astronomer watching the developments in Perseus now sees and describes to us the destruction of the world (of which the 'heavens are seen to melt with ferment and heat,') and the simultaneous beginning of the new heaven and the new world, the process of the sequence prophesied alike by Laplace and the inspired seer," thereby at once "confirming an old, and giving a new and more exact, revelation."[18]

Recounting the evolution of mankind's exploitation of energy and predicting the inevitable future "perfection" of energy utilization, Thurston predicted that "man, guided by nature, should be able, in a comparatively brief period, to reach the same end." "Directing every energy precisely to the accomplishment of its prescribed purpose, applying every substance in its right place and in the right manner in his constructions, and bending every law to his aid in the building of a world, he may profit in maximum degree by every force, energy and substance, by all material and all spiritual laws and phenomena, by all opportunities of advancing himself to loftier and loftier planes, perfecting himself...." If the mechanical means had changed over time, the perfectionist ends of the religion of technology remained very much the same.[19]

Francis Bacon "saw about him a world awaiting deliverance," declared Ralph Flanders, another ASME president, and this destiny would now, finally, be achieved by the engineer, the "New Man" of the "New Age." The advance of technology would surely bring about "The Great Awakening," as Albert Merrill, pioneer aeronautical engineer, entitled his own technological utopia. Thomas A. Edison, the American embodiment of this "New Age," viewed himself as a disciple of Michael Faraday, and not just in his researches on electricity. Like the Sandemanian Faraday, Edison, the ultimate utilitarian, also aspired to know more transcendent terrain. Throughout his career, he was "careful to stress that his ideas did not contradict the Bible." "I am not an atheist," he declared; "the greatest monument of all times was the Cross of Calvary. It has had a greater effect on more people for a longer time than any other thing erected by man." "He had faith in providence," one close friend recalled, and his technological labors only heightened it. "When you see everything that happens in the world of science and in the working of the universe, you cannot deny that there is a 'captain on the bridge,'" Edison told him. "The existence of . . . God, in my mind," he later declared, "can almost be proved by chemistry." His own subdued conventional religiosity was demonstrably surpassed by that of his wife, Mina, who had a church built adjacent to their property in Fort Myers, Florida, the Thomas A. Edison Congregational Church, and later joined the religiously inspired American movement for "Moral Re-Armament." As her husband had brought light into people's homes, she explained, she wanted to flood their hearts and minds "with God's illumination."[20]

But there was also a less conventional aspect to Edison's explorations of eternity. Following a path pursued earlier by other scientists, such as Leibniz and Swedenborg, Edison ventured into more mystical spiritual exploration and experimentation. According to one biographer, he betrayed a credulous and "persistent inclination to explore the unquantifiable aspects of reality"—"realms beyond," life after death—and "visited and revisited these mystical spheres periodically throughout his life, up until the very end." And at the very end, awaking from a coma for one final moment, he uttered his last words: "It is very beautiful over there."[21]

The robust spirit of the religion of technology in America was perhaps best expressed and most effectively popularized by the technological-utopian writings of the American socialist Edward Bellamy. Born of a Baptist minister father and a zealously religious mother, Bellamy was firmly rooted in the fervent evangelical tradition of rural New England. At the same time, his works reflect the liberal attitudes of his father and grandfather (who was forced from his pulpit because of his Freemasonry), and the unmistakable influence of Auguste Comte. Bellamy's writings thus resound with the familiar refrains of redemption, of the divinely destined recovery of mankind's lost perfection.[22]

In his first book, *The Religion of Solidarity*, Bellamy described the "tendency of the human soul to a more perfect realization of its solidarity with the universe, by the development of instincts partly or wholly latent." "In the soul is a depth of divine despair over the insufficiency of its existence . . . and a passionate dream of immortality," Bellamy declared. "The half-conscious God that is man is called to recognize his divine parts." In a later edition of the book, Bellamy confirmed that it "represents the germ of what has been ever since my philosophy of life" and expressed the wish that it be read to him on his deathbed.[23]

In 1888, Bellamy published his enormously popular and influential utopian novel, *Looking Backward*, which immediately became one of the best-selling books of the nineteenth century and inspired the efforts of generations of social reformers. Bellamy's futuristic portrayal of America at the turn of the second Christian millennium was perhaps the quintessential "product of America's peak of faith in technology," as historian Howard P. Segal has emphasized. "The United States of the year 2000 is very much a technological utopia: an allegedly ideal society not simply dependent upon tools and machines, or even worshipful of them, but outright modeled after them. . . . The purposeful, positive use of technology—from improved fac-

tories and offices to new highways and electric lighting systems to innovative pneumatic tubes, electronic broadcasts, and credit cards—is,
in fact, critical to the predicted transformation of the United States
from a living hell into a heaven on earth."[24]

The time-traveling protagonist of the novel, Julian West, is symbolically resurrected from his nineteenth-century subterranean refuge
into a paradisiacal future age, a late-twentieth-century socialist society which signifies "the greatness of the world's salvation" and "deliverance." The millenarian significance of the new age is articulated
most forcefully in the form of a sermon delivered by Mr. Barton.
"Humanity is proving the divinity within it," Barton proclaims, with
"a vista of progress whose end, for very excess of light, still dazzles
us." As the new day dawned, Barton recounted, "for the first time
since the Creation every man stood up before God. . . . It was for the
first time possible to see what unperverted human nature really was
like." In the language of Comte and more distant predecessors, Barton described how mankind "had sprung back to its normal upright-
ness," revealing its "god-like aspirations [and] impulses . . . images of
God indeed."[25]

The new age "may be regarded as a species of second birth of
the race, . . . a new phase of spiritual development," Barton declared,
echoing Joachim of Fiore. "For twofold is the return of man to God
'who is our home.' The return of the individual by the way of death,
and the return of the race by the fulfillment of the evolution, when
the divine secret hidden in the germ shall be perfectly unfolded. With
a tear for the dark past, turn we then to the dazzling future, and, veil-
ing our eyes, press forward. The long and weary winter of the race is
ended. Its summer has begun. Humanity has burst the chrysalis. The
heavens are before it."[26]

In the years following the publication of *Looking Backward*,
Bellamy tempered his enthusiasm for technological deliverance.
Thrust into political engagement, he was forced to reflect further and
revise and refine his analysis. In a brilliant chapter of his masterwork
Equality, which was published a decade later, shortly before his
death, Bellamy sought to explain why the most dramatic advances in
technological development had not only failed to better the lives of

most people but had actually contributed to their misery—"a fact which to our view absolutely overshadows all other features of the economic situation." Furthermore, he tried to fathom why there had been so little serious concern about this glaring paradox. He observed with astonishment that the lack of any apparent benefit did not at all diminish the extravagant popular enchantment with new inventions, and he recognized in this irrational behavior a deep-seated cultural compulsion. "This craze for more and more and ever greater and wider inventions for economic purposes, coupled with apparent complete indifference as to whether mankind derived any ultimate benefit from them or not," he surmised, "can only be understood by regarding it as one of those strange epidemics of insane excitement which have been known to affect whole populations at certain periods, especially of the middle ages. Rational explanation it has none."[27]

PART II ——————————————

TECHNOLOGIES OF
TRANSCENDENCE

CHAPTER 8

ARMAGEDDON:

ATOMIC WEAPONS

For a relatively brief period, as it now seems in retrospect, the other-worldly apocalyptic aspect of Christian mythology gave way to more earthly expression, epitomized by Edward Bellamy's socialist utopia. Whereas Bellamy at least retained the idea of a millennial kingdom, most of his contemporaries opted instead for a more secular view of mankind's unending evolution. Thus, for a century or so, as Perry Miller observed, "an image of infinite progress bit by bit blotted out the ancient expectation." Ideologically rooted in the science-based progressivism of the eighteenth-century Enlightenment, this optimistic open-ended outlook was fueled by a remarkable and seemingly endless succession of technological and industrial advances, as well as by new evolutionist theories of gradual but steady biological and social development. (Among millenarians, grown impatient by the failure of countless predictions of the apocalypse, this progressivist outlook took the form of postmillenarianism, a belief that the millennium had already begun and that Christ would return only after mankind had by its own hands created his earthly kingdom.) But beneath the surface of this willfully hopeful modern vision, the other-worldly millenarian mentality remained intact at the core of Western culture.[1]

Masked by a secular vocabulary and now largely unconscious, the old religious themes nevertheless continued subtly to inform Western projects and perceptions. In times of crisis, which momentarily shook men from their progressivist complacency, these themes once again became explicit. Thus, after nearly a century of great-power-brokered peace, the two world wars sparked a renewal of apocalyptic thinking, as did the Nazis' explicitly millenarian vision of a thousand-year Third Reich. At the same time, a number of horrific war-making innovations—from aerial bombardment and chemical warfare to nuclear weapons—seemed to signal, in their very potential for death and destruction, the pacific promise of new beginnings. And here especially the apocalyptic religion of technology re-emerged with a vengeance, making a mockery of comparatively thin-sounding hymns to progress. For such new technologies promised not mere incremental steps toward perfection, but history-shattering, transcendent leaps of doom and deliverance. The calamitous events and diabolical designs of the twentieth century weakened the fashionable faith of modernity by recalling more ancient imaginings. As an already precarious belief in progress was "blasted by the atomic flash," millenarian dreams and nightmares returned anew to stir and haunt Western consciousness, and to color the technological imagination.[2]

The hallmark technologies of the period thus came to reflect the anxieties and anticipations of an earlier age. As Lewis Mumford wrote, "Fantasies of the seventeenth century have often proved closer to our own twentieth century realities than the more humanly fruitful but relatively pedestrian enterprises of eighteenth and nineteenth century industry." This latest apocalyptic awakening began among the atomic technologists themselves, who first knew of and fully recognized the awesome significance of the potent new force in the making. As Roger Bacon seven centuries before had urged the pope to develop and exploit new inventions lest the Antichrist seize them for evil advantage, so the atomic physicists alerted their political masters to the new ominous potential, warning them to steal a lead on their enemies in order to ensure that this power would be used by the forces of good rather than those of evil. In the right hands, they believed, the use of this technology could be a blessing, a means of salvation by which mankind would be transported from a dismal history of division and strife into a new era of global cooperation and peace. As those who discovered and could unleash nuclear energy on the world, the atomic scientists and engineers viewed themselves, in an almost divine light, as the veritable saviors of mankind.[3]

Leo Szilard, the engineer-turned-physicist who initially conceived of the possibility of a nuclear chain reaction and first sounded the alarm about its lethal and liberatory significance, saw himself as the leader of a new breed of spiritual men destined to bring order and light to the world. Szilard viewed the release of atomic energy as a means of transcending not only earthly travails but also the earth itself—"If I wanted to contribute something to save mankind," he wrote, "then I would probably go into nuclear physics, because only through the liberation of atomic energy could we obtain the means which would enable man not only to leave the earth but to leave the solar system." Throughout his life, beginning in the mid-1920s, he dreamed of forming yet a new brotherhood, Der Bund, in the Rosicrucian tradition, which he described as "a closely knit group of people whose inner bond is pervaded by a religious and scientific spirit." "Through education in close association," Szilard wrote, "we could create a spiritual leadership class with inner cohesion which would renew itself on its own."[4]

In this spirit, and inspired by the world-renewing promise of what Ernest Rutherford, the father of nuclear physics, described as "a newer alchemy"—the nuclear transmutation of the elements—Szilard began in 1930 to organize a group of acquaintances, most of them young physicists, into a working association. He subsequently labored to aid many in their escape from Hitler's Europe, and ultimately initiated the effort that led to the creation of the Manhattan Project, the largest engineering project in history, and the formation of the fateful fraternity of Los Alamos.[5]

At first cautiously couched in understated diplomatic rhetoric, the apocalyptic enthusiasms of the nuclear pioneers became explicit when the full fury of their primordial force was finally released. The first atomic explosion, the Trinity test in Alamogordo, New Mexico, on July 16, 1945, was a secret affair, witnessed and known only to a select corps of technical, military, and political personnel. Chief among these was Robert Oppenheimer, who as director of the Los Alamos laboratory was the administrative, intellectual, and spiritual guide of the atomic brethren. Oppenheimer was immediately responsible for the success of the atomic-bomb project. Like many of his colleagues a Jew, he provided the explicitly religious, and decidedly Christian, code name for this epoch-making event: Trinity.

"The first man-made nuclear explosion would be a historic event and its designation a name that history might remember," one historian of the atomic bomb noted; "Oppenheimer coded the test and the test site Trinity." The religious significance of the term was intended. Two decades later, Oppenheimer explained to General Leslie Groves, who had been the military commander of the Manhattan Project, what he was thinking when he christened it. "Why I chose that name is not clear, but I know what thoughts were in my mind. There is a poem by John Donne, written just before his death, which I know and love. From it a quotation: 'As West and East / In all flatt Maps—and I am one—are one, / So death doth touch the Resurrection.'" Oppenheimer also pointed out that another of his favorite Donne poems opens with the line "Batter my heart, three person'd God." "Beyond this, I have no clues whatever," he concluded.[6]

Thus the leader of the effort to build the atomic bomb had on his mind when he named the first bellow of the beast the redemptive reveries of a seventeenth-century cleric, a contemporary of Francis Bacon. The first poem from which he quotes, which he loved and memorized ("Hymne to God My God, in My Sicknesse"), goes on in an equally familiar vein:

> We thinke that *Paradise* and *Calvarie*
> *Christs* Crosse and *Adams* tree, stood in one place;
> Looke Lord and find both *Adams* met in me;
> As the first *Adams* sweat surrounds my face,
> May the last *Adams* blood my soule embrace."[7]

"Both *Adams* met in me": the first man, image of God, fallen favorite of creation, and Christ, the last Adam reborn from death to inspire and lead the second creation, of man's redemption. The significance of the poem is clear enough, that dying leads to death but also to the prospect of a more blessed renewal. So the bomb for Oppenheimer, as for many of his colleagues, especially Szilard and Niels Bohr, signaled a beginning as well as an end, "a weapon of death that might also redeem mankind." Oppenheimer "cherished the complementary compensation of knowing that the hard riddle the bomb would pose had two outcomes, one of them transcendent." But the Adamic emphasis in Donne's poem indicates also the deeper meaning of such transcendence: the restoration of Adam's original perfection and divine-likeness. Mesmerized by their spellbinding achievement, the atomic pioneers behaved, like so many earlier self-anointed saints, as if they had themselves already advanced toward this recovery of divine-likeness, as much Redeemers as redeemed. As one of them later reflected, their development of the bomb gave them "the illusion of ultimate and illimitable power, like being God."[8]

The explosion of the first bomb prompted an awestruck apocalyptic reaction. "I am become death, destroyer of worlds," said Oppenheimer, quoting the *Bhagavad Gita*. "In the last millisecond of the earth's existence the last man will see something very similar to what we have [seen]," said chemist George Kistiakowsky, who had prepared the explosives for the device. William Laurence of *The New York Times*, who served as the official military reporter at Los Alamos, gave voice to the inarticulate. Unconsciously echoing a thousand years of elite millenarian expectation, he likened this new creation to the old, comparing the work of man with the work of God. "This rising supersun seemed to me the symbol of the dawn of a new era," he exulted. In the same breath, he noted that "one felt as though he had been privileged to witness the Birth of the World; . . . if the first man could have been present at the moment of Creation when God said 'Let there be light' he might have seen something very similar to what we have seen." General Thomas Farrell, Groves's deputy at Los Alamos, underscored the connection, observing that the Trinity test had unleashed forces "heretofore reserved to the Almighty."[9]

The secret of Los Alamos was made public at Hiroshima. "After centuries of calculation, the date and moment became precise: it was 0815 hours ... on 6 August, 1945, and the place was not Rome at all," Perry Miller wrote of Hiroshima's millenarian significance. "The latest contribution to the literature of the apocalypse marks an innovation: the narrative for the first time becomes historical." The descriptions of the explosion bore uncanny resemblance to earlier stylized descriptions of the apocalypse: a light, then a blast, then the flames from which there was no hiding place. "Which at midnight brake forth a Light / which turn'd the night to day, / and speedily an hideous cry / Did all the world dismay," wrote Michael Wigglesworth in 1662. "And if the wickedness of the old world, when men began to multiply on the earth, called for the destruction of the world by a deluge of waters," warned Jonathan Edwards a century later, "this wickedness will as much call for its destruction by a deluge of fire." According to the official United States Bombing Survey description of the Hiroshima explosion, "An intense flash was observed first, as though a large amount of magnesium had been ignited, and the scene grew hazy with smoke. At the same time at the center of the explosion, and a short while later in other areas, a tremendous roaring sound was heard and a crushing blast wave and intense heat were felt. ... The atomic bomb shattered the normal fabric of community life. ..."[10]

Hiroshima, a stark reminder of the impermanence of progress and the contingency of history, provoked what historian Paul Boyer described as an "atom-induced revival of eschatological thinking," and a "mood of approaching apocalypse." "This atomic bomb is the Second Coming in Wrath," exclaimed Winston Churchill. "The atomic bomb is the good news of damnation," declared Robert Hutchins. Theologians seized the moment. "For generations the moral obligation of Christians 'to make preparation for the world's end' had been ignored or relegated to the subconscious," Wesner Fallow of the Andover Newton Theological Seminary intoned in 1946, but now "eschatology confounds us at the very center of consciousness." The Methodist leader Ernest Fremont Tittle warned that same year that "we have now, apparently, to reckon with the possibility of a speedy end to man's life on earth. . . . What is new in the present situation is not the possibility of a last generation but the possibility . . . that *ours* may be the last generation!" The atomic bomb, he noted, reminded Christians that history "is not limited to this passing world but will have its consummation in the eternal Kingdom of God."[11]

Immediately following the successful test of the Soviet bomb in 1949, renewed predictions of global annihilation spawned a revival of evangelical expectation. The Baptist evangelist Billy Graham, the most successful revivalist of the new apocalyptic vision, assailed the Antichrist of godless communism and warned the wayward of the imminence of Armageddon, the mythic battle between the forces of Jerusalem and the Antichrist briefly mentioned in Revelation. The world is "moving now very rapidly toward its Armageddon," declared Graham; "the present generation of young people may be the last generation in history."[12]

In subsequent decades, with the precarious Cold War nuclear stalemate providing the common backdrop for all the world's adventures, a succession of remarkably successful preachers, armed especially with the prophecies of Daniel and Revelation, repeated fervent calls for repentance in the face of inevitable doom. Some, especially the so-called born-again fundamentalists of the dispensationalist persuasion, actually welcomed the nuclear holocaust as Armageddon—that is, as a fulfillment of prophecy and as a sign that the millennium was at hand. Often they preached that they and their flock, the faithful remnant, would be spared the horrors of this fateful conflagration. As the battle began, they would miraculously ascend to join Christ in the air (the idea of the Rapture taken from 1 Thessalonians 4:16–17) to return later and reign with him for a thousand years. The fundamentalist preacher Jerry Falwell pointedly identified Armageddon with nuclear war and encouraged his followers eagerly to embrace the prospect as a promise of deliverance. "I believe there will be some nuclear holocaust on this earth," said Falwell. "There will be one last skirmish and then God will dispose of this Cosmos. The Scripture tells us in Revelation, chapters 21 and 22, God will destroy this earth—the heavens and the earth. And Peter says in his writings that the destruction will mount as with a fervent heat or a mighty explosion." "You know why I'm not worried?" Falwell asked his followers. "I ain't gonna be here."[13]

If the nuclear Cold War era fueled a revival of such ancient imaginings, this in turn provided a fatalistic framework for the further development of nuclear weaponry. It gave cosmic sanction, for example, to the work of those who assembled all of the American nuclear weapons at the Pantex plant in Amarillo, Texas. As novelist A. G. Mojtabai has shown, these people produced implements of death in just such a fatalistic spirit. "I think that it's just the fulfillment of prophecy in the Bible that man will become so corrupt, so evil and vile . . . that, one of these days . . . God is going to run out of patience," one Pantex inspector told Mojtabai. "I think that the things are in God's timetable and these times are going to happen and ain't a whole lot we can do," added a Pantex engineering technician. "To me, a Christian is worth . . . the chances that we take of a nuclear holocaust."[14]

Pantex employees were strongly encouraged in such belief by their local evangelical clergy. "There's coming a time when every Gentile nation on the face of this earth shall gather their forces . . . to fight against the army of God in what is called the battle of Armageddon," Charles Jones of the Second Baptist Church of Amarillo confidently declared, quoting from 2 Peter 3 ("the heavens shall pass away with a great noise, and the elements shall melt with fervent heat, the earth also and the works that are therein shall be burned up"). But, he added, "God's people will not be in that final battle—they'll be caught up in a chariot of clouds to meet the Lord in the air and so shall we ever be with the Lord." "God's going to do something to the earth. Man, I'm not going to be here. I'm going to be in glory with Jesus cause I've been saved. I don't want to be here. I want to be in glory." The same message was offered by Royce Elms of the First United Pentecostal Church of Amarillo. "You say, brother Elms, are you talking about nuclear holocaust for the USA. Do you mean to tell me that we're going to be the victim of a terrifying nuclear attack? Absolutely! It is ordained in God's Word beyond any shadow of a doubt." But, he reassured his followers, "my church, my people, you're not gonna be there when the bomb starts falling. I'm gonna take you outta here!"[15]

The revived millenarianism of the nuclear age was based upon a renewed belief in both inevitable technological destiny and deliverance, by way, ultimately, of an atomic Armageddon. Here the fatalistic fascination with final things focused squarely upon technological developments, viewed through the lens of what psychologist Robert Jay Lifton described as a collective psychological "blending of ultimate destruction and human redemption." The disintegration of the atom and the fateful release of the fundamental energy of creation would, through its furious force, at last reintegrate humanity with its creator. "If the bomb dropped today, it wouldn't bother me one bit," a Pentecostal preacher declared. "The whole world would come into a knowledge of Jesus Christ and we would have peace."[16]

Among the technical elite still involved in the development of the nuclear technology, the atom bomb induced a mood of approaching apocalypse, more subdued and secular perhaps, but no less intense. For them probably more than anyone else, the imperative of technological development defined their lives, fostering an almost fetishistic faith in technological destiny and fueling their own desperate dreams of technological transcendence. Also for them, as Lifton observed, "nuclearism" was a distinctly "millenarian ideology," and hence the prospect of nuclear doom was more comfortingly construed as the promise of nuclear deliverance. "Secular and religious Armageddon images tend to merge in many minds," Lifton noted. "Armageddonist imagery," for example, "can also be held by those close to the weapons, and may include impulses to purge the world of its evil by means of nuclear holocaust." Driven by their own technological compulsions and delights, a "totalistic" paranoid perception of the enemy as the embodiment of evil, and a belief in their own unique power and destiny to save the world, these "secular Armageddonists" nevertheless "renounce responsibility for the holocaust they anticipate and may press toward bringing about," and "may view nuclear holocaust as an inevitable outcome of our time and technology which is pointless to resist."[17]

Lifton's portrayal of the "secular Armageddonist" mentality accurately describes the millenarian world-view of the technologists who designed the later generations of nuclear weapons, from Edward Teller and his associates on the H-bomb project to Teller's direct disciples and descendants at the Lawrence Livermore Laboratory. All presupposed the diabolical designs of the Soviet Antichrist and the inevitability of a final nuclear showdown. For half a century Teller himself unswervingly displayed what has been described as a "religious dedication to thermonuclear weapons," while his followers in the arms race derived "a very large part of their self-esteem from their participation in what they believe to be an essential—even a holy—cause," as Livermore's first director, Herbert York, described them.[18]

Szilard's original Rosicrucian vision of Der Bund was realized at Livermore, where a tight-knit group of elite technologists was carefully assembled to carry on the spirit of the Manhattan Project. Isolated from the world by high security as well as by a peculiar set of customs, shared experience, and private language, theirs was "a very closed society," akin to a monastery. Recruitment was carried out by Livermore personnel, under the leadership of Teller protégé and Livermore abbot Lowell Wood, through a deliberate process of selection from among the cream of the crop of the nation's brightest young technical talent. (The chief instrument of selection was the private Hertz Foundation, established by the car-rental company's founder and directed by Teller and other nuclear zealots, which lured prospective—and often unsuspecting—young weaponeers by means of lucrative internships and stipends. The sites of recruitment were the nation's premier engineering schools, particularly MIT and Caltech.) Once within the circumscribed confines of the weapons laboratory, the young recruits—referred to affectionately as Teller's "sons" and "grandsons"—were habituated to the norms of the community by persuasion, discipline, and new bonds of allegiance, as well as by high-powered technological competition and enchantment. Before long, they invariably adopted the apocalyptic outlook of nuclear-age spiritual men, their conversations often tending toward "a favorite topic—global extinction."[19]

Livermore's Lowell Wood, who came under Teller's spell while still an undergraduate, seconded his mentor's convictions. Like other nuclear-weapon designers, Wood believed that his work was essential for the salvation of human society—"weapons of life"—because they deterred nuclear conflict. At the same time, however, he was convinced of the inevitability of a deliberate Soviet pre-emptive nuclear attack. "Someday—maybe out of the clear, blue sky, because that's the way they're postured, that's the way they're wired; BAM! It will be all over." One of his recruits, Caltech graduate Larry West, likewise viewed himself as contributing to mankind's salvation. A designer of supercomputers as well as nuclear weapons, West told *New York Times* science writer William Broad, "I consider computers to be as much a weapon as nuclear warheads are. They have as much importance to the salvation of society." West too was resigned to the inevitability of a nuclear showdown. "Maybe the best way to eliminate arms is just to throw the missiles into the sea, totally disarm, and shake hands with the Russians. I've thought about that. But I don't believe it's a real possibility." MIT recruit Peter Hagelstein, young inventor of the nuclear X-ray laser (centerpiece of the Strategic Defense Initiative), never really wanted to get into the weapons-design business but fell prey to intense competition with colleagues and the allure of technically challenging problems. Though half convinced that he was working only on "defensive" nuclear weapons, he well understood that defense and offense in nuclear confrontation were fundamentally indistinguishable, and that both would invariably impel the world toward inevitable disaster. "I'm more or less convinced," he told Broad, "that one of these days we'll have World War Three, or whatever."[20]

Or whatever. The apocalyptic outlook of the weapons-designers is, in essence, no different from that of the evangelist: the expectation of inevitable doom. And here too anticipation of annihilation is "blended" with a belief in salvation. For the weapons-designers, the bomb is a means not only of destruction but also of deterrence, defense, and deliverance. If nuclear weaponry does not deter attack, it might defend at least some of the species from earthly extinction. And if that too fails, it might be used instead to propel a privileged few scientific saints to safety among the stars. For all their claims of building bombs to avoid disaster, at least some in the nuclear community were hedging their bets by seeking yet another form of technological transcendence, their own technical version of the Rapture: nuclear-powered spaceflight.

This was the early dream of Leo Szilard, inspired by the H. G. Wells novel *The World Set Free,* which envisioned just such an atomic escape from atomic catastrophe. It was also the determined fantasy of Stanislaw Ulam, co-inventor with Teller of the hydrogen bomb. Physicist Freeman Dyson, author of a 1958 "Space Traveler's Manifesto," collaborated on the development of nuclear weapons in the hope of securing sufficient power for his imagined starship, mankind's ultimate means of achieving a universal and immortal existence. At Livermore, this particular idea of technological transcendence was kept alive by Rod Hyde, group leader for nuclear-weapons development, inventor of the nuclear-bomb-pumped gamma-ray laser, and designer of his own nuclear-bomb-propelled starship. "What I want more than anything is to get the human race into space," Hyde told Broad. "It's the future. If you stay down here some disaster is going to strike and you're going to be wiped out. . . . My idea of the future is to get off into space." True to apocalyptic tradition, as Perry Miller explained, when the end is approaching "the saints will know—as indeed they know even now—that they must ascend into Heaven."[21]

CHAPTER 9

THE ASCENT OF THE SAINTS: SPACE EXPLORATION

What today we call space used to be known as heaven. From its earliest expressions, the enchantment of spaceflight was fundamentally tied to the other-worldly prospect of heavenly ascent. Just ten years after the death of Columbus, who, in quest of the terrestrial paradise, became the greatest of earthbound explorers, the Italian Renaissance poet Ludovico Ariosto envisioned the next, extraterrestrial, step in mankind's return journey to God. In his great epic, *Orlando Furioso*, published in 1516, Ariosto imagined a new means of escape from a fallen world. Astolfo, exploring the earth in a time of troubles, discovers the terrestrial paradise on a mountaintop. There he encounters St. John the Evangelist, who proposes that Astolfo continue on— "a flight more daring take, to yonder Moon." A century later, in 1611, the millenarian mystic Tommaso Campanella wrote to Galileo explaining how he had "read new meaning into a familiar verse, 'and I saw a new Heaven and a new earth'—namely, that the moon and the planets were inhabited." In his later *Apologia pro Galileo* he suggested the possibility that paradise was not really terrestrial at all, but lay on the moon, which was situated high enough above the earth to have been spared the deluge of the flood. The moon must have a moderate climate, Campanella argued, because its Edenic inhabitants, "not infected with Adam's sin," went naked.[1]

Campanella's equally devout contemporary Johannes Kepler had similar imaginings. In 1609, Kepler wrote of his celebrated dream about a voyage to the moon, the same year he published the laws of planetary motion which three and a half centuries later would guide other Christians to an actual lunar landing. As a youth, Kepler attended the convent school at Maulbonn, haunted by the spirit of Dr. Faustus. He remained throughout his life deeply religious; like Newton, he devoted considerable energies to deciphering biblical chronologies and maintained an earnest belief in the possibility of resurrection. Having originally trained in theology in preparation for the clergy, he devoted his scientific labors above all to the "glorification of God" and identified astronomers as "priests of the highest God" (he was closely associated with the Rosicrucians). "There is nothing I want to find out and long to know with greater urgency than this," he wrote a friend, "can I find God, whom I can almost grasp with my own hands in looking at the universe, also in myself?" In his heroic efforts to comprehend the celestial order, he sought somehow to reconcile "the heavenly mind of man to this dusty exile of our earthly home." "Should . . . the kind of creator who brought forth nature out of nothing . . . deprive the spirit of man, the master of creation and the Lord's own image, of every heavenly delight?"[2]

Like so many of his contemporaries—and descendants—he despaired about the terrors of his war-ravaged times and sought refuge in the stars. "Would it not be excellent to describe the cyclopic mores of our time in vivid colors, but in doing so—to be on the safe side—to leave this earth and go to the moon?" Kepler wrote his friend. "As we are driven from this earth, [my astronomy of the moon] will be useful to us as a viaticum on our wandering to the moon." Having earlier speculated about the planetary motion of the earth, despite its static appearance to its inhabitants, by imagining how the earth would appear from the perspective of an observer on the moon, Kepler in his dream wondered how such an observer might get there. In his *Som-*

nium, the account of his dream, Kepler imagined an earthling soaring high above the mountains into space as if shot from a cannon. As Columbus and other intrepid explorers had traversed the vast oceans, he wrote to Galileo, "let us create vessels and sails adapted to the heavenly ether." "There will be plenty of people unafraid of the empty void. In the meantime, we shall prepare, for the brave sky-travellers, maps of the celestial bodies—I shall do it for the moon, you, Galileo, for Jupiter."[3]

Kepler's *Somnium* was a familiar reference for all later writers of cosmic voyages. In 1638, John Wilkins, one of the founding fathers of the Royal Society, wrote his *Discourse Concerning the Discovery of a New World in the Moon,* drawing upon the earlier speculations of both Campanella and Kepler. He noted that many had "affirmed that Paradise was in a high elevated place, which some have conceived could be no where but in the Moon," and that "Kepler doubts not, but that so soon as the art of flying is found out, some of their Nation will make one of the first colonies that shall inhabit that other world." Emboldened by "a contempt for these earthly things," exulted Wilkins, "how happy shall they be, that are first successful in this attempt." "All this place wherein we war . . . is but a point far less than any of those small stars, that at this distance are scarce discernible, which when the soul does seriously meditate upon, it will begin to despise the narrowness of its present habitation, and think of providing for itself a mansion in those wider spaces above, such as may be more agreeable to the noblenesse and divinity of its nature."[4]

In like spirit, the Royal Society's Joseph Glanvill also earnestly entertained the prospect of an airborne "voyage to the Southern unknown tracts, yea possibly to the moon," as did such later visionaries as Bernard Fontanelle and Christian Huygens. But perhaps the most influential interplanetary prophet of them all was the nineteenth-century inventor of science fiction, Jules Verne, who had also been inspired by Kepler. The product of a stern, pious, orthodox Catholic upbringing, Verne too was a devout Christian, and his inspirational scientific fantasies resound with familiar religious strains.[5]

In his 1865 novel *From the Earth to the Moon,* Verne told a tale about a lunar voyage that excited the imagination of nearly all of the pioneers of actual spaceflight. Although the means of travel stretched credulity—the space capsule containing the lunar explorers was shot from a large cannon—Verne accurately anticipated many features of what would become the real thing, including the location of the launch site (central Florida), the experience of weightlessness, the shape of the capsule, the use of rockets to alter orbit, and the splash-down at sea. In addition, Verne chose as his protagonists in the re-markable adventure military men whose Civil War experience as artillerists had equipped them for this technological leap. They were members of the Gun Club, demobilized soldiers seeking an outlet for their frustrated aggressions and some use for their ballistic hard-ware and talents. His description of these men eerily anticipates the weapons-delivery-system designers who became the pioneers of space travel. "The estimation in which these gentlemen were held . . . was proportional to the masses of their guns, and in the direct ratio of the square of the distances attained by their projectiles. . . . It is obvious that the sole preoccupation of this society was the destruction of hu-manity in a spirit of philanthropy and the perfecting of weapons of war considered as instruments of a civilizing mission." Verne referred to the group as "Exterminating Angels."[6]

The religious overtones of this novel are suggested in the use of allusions to the "Ascension" and biblical descriptions of the origins of the universe, as well as in the "religious enthusiasm" of Barbicane, leader of the Gun Club and initiator of its extraterrestrial undertak-ing. "It is perhaps reserved for us to become the Columbus of this un-known world," he declares (the cannon is called the "Columbiad"). "If we die, the result of our travels will be magnificently spread. It is His own secret that God will tell us! In the other life the soul will want to know nothing, either of machines or engines! It will be iden-tified with eternal wisdom."[7]

The implied religiosity of all this technological heroism becomes more explicit in Verne's sequel to his moon voyage, *Sans Dessus Dessous* (*Upside Down*), which is the fulfillment of the Gun Club fantasy. Here the half-crazed artillerists embark on another pro-methean project, this time using their ballistic genius to design an even bigger cannon, the firing of which would be used to correct the tilt in the earth's axis, thereby eliminating extremes in climate. As one Verne scholar noted, "The new order to be ushered in by the shift in axis is represented as a wholly new start, a new beginning, comparable to the divine act of creation itself, . . . a 're-origination.'" "The final version of the consequences of the blast is strictly apocalyptic in tone and content." Indeed, Verne borrows heavily from the Book of Revelation, referring to the Antichrist and end-of-the-world prophecy, and thus "offers an updated, rigorously scientific version of Revelations, a technologically engineered Day of Judgement."[8]

The full religious message in Verne's novels is finally revealed in his last work, his "testament," a short story entitled "The Eternal Adam." Here the threatened end of the world, avoided in *Sans Dessus Dessous*, at last takes place. "Earth is now identical and coterminous with the Empire. The dream of Empire has finally been realized. . . . Humanity is now ready for the Truth." This universal imperium "recalls the Garden of Eden." Watered by four rivers, it is a "man-made paradise, a mirror image of Genesis, the end of evolution." "Scarcely has man appeared on earth than he immediately begins and unceasingly continues his ascent," wrote Verne. "Slowly but surely, he approaches his end, which is the perfect knowledge and absolute domination of the universe."[9]

Why do men climb the tallest mountains and voyage to the moon? asked science-fiction author Ray Bradbury, writing about Verne's influence. The reason "Verne implies is, we go there because we are nearer the stars, and if we reach the stars, one day we will be immortal." "We are all, in one way or another, the children of Jules Verne," Bradbury wrote. "His name never stops. At aerospace or NASA gatherings, Verne is the verb that moves us to Space. . . . Without Verne there is a strong possibility we would never have romanced ourselves to the Moon." Not only did Verne influence later science-fiction writers, notably H. G. Wells, but he inspired the rocket pioneers who devised the actual mechanical means by which mankind might finally be freed from its "dusty exile" on earth.[10]

Early rocket development paralleled that of powered flight, which, as historian Michael Sherry has noted, was itself suggestive of such emancipation. "Never viewed solely as a weapon, the airplane was the instrument of flight, of a whole new dimension in human activity. Therefore it was uniquely capable of stimulating fantasies of peacetime possibilities for lifting worldly burdens, transforming man's sense of time and space, transcending geography, knitting together nations and peoples, releasing humankind from its biological limits. Flight also resonated with the deepest impulses and symbols of religious and particularly Christian mythology—nothing less than Christ's ascension. Its realization then served as a powerful metaphor for heavenly aspirations and even, among the literal-minded, as the palpable vehicle for achieving them."[11]

The strict sectarian religious upbringing of Orville and Wilbur Wright perhaps rendered them the perfect representatives of mankind's first steps toward heavenly ascent. Their father, an archconservative bishop of the Church of the United Brethren in Christ, was editor of the sect's organ, *The Religious Telescope*, and Wilbur, the older brother, who first conceived of the idea of the airplane, worked closely with his father and wrote some influential church tracts. Later the Wrights forbade their employees to participate in any flights on Sundays. The austere celibate lives of these elusive, paranoid recluses suggests perhaps unspoken other-worldly aspirations, much like those of their spiritual brethren the rocket pioneers Konstantin Tsiolkovsky and Robert Goddard.[12]

Tsiolkovsky, a Russian schoolteacher, is generally credited with having first laid the scientific foundation for modern rocketry (and for later Soviet space efforts) before the end of the nineteenth century. Rendered partially deaf by scarlet fever at the age of ten, which "made me a victim of ridicule," Tsiolkovsky turned inward. "Since childhood, partial deafness resulted in a total ignorance of the ways of everyday life, and therefore a life of 'connections,'" he wrote in his autobiography. "This handicap estranged me from people and prompted me to read, concentrate, and dream. . . . There was a desire to do something big, heroic. . . . All my life consisted of meditation, calculations, and experimental work." Inspired by Jules Verne, he grappled with the technical challenge of rocketry. "Probably the first seeds of the idea were sown by that great fantastic author Jules

Verne," Tsiolkovsky recalled. "He directed my thought along certain channels, then came a desire, and, after that, the work of the mind." At sixteen, he thought that he had discovered in centrifugal force the means of "flight in cosmic space." "I still remember that night," he fondly remembered, "and even now, fifty years later, I sometimes dream about rising in my machine toward the stars and feeling the same exaltation."[13]

Tsiolkovsky's interest in spaceflight was inspired also by his early association with the influential Russian mystic Nikolai Federov, who became his spiritual father. Federov played a crucial role in Tsiolkovsky's formative years, rescuing him from suicidal despair and teaching him that mankind's ultimate destiny included, and indeed required, "conquest of the cosmos." Federov combined the ideals of Russian Orthodoxy, the Russian aristocracy, and the Russian peasant commune into a doctrine of what he called "the Common Task," the unification of all humanity and the "removal of all the obstacles that prevented the evolution of man's humanity toward its last stage, the stage of self-creation, immortality, and God-likeness." In Federov's vision, "mankind's purpose in Creation was the transformation of our mortal universe into an immortal cosmos," which ended with mankind's "consubstantiality with God." This transformation, which entailed the reconstitution of the bodies of past humans, demanded mankind's complete mastery and control over the universe, including space. Moved by this vision of his spiritual master, Tsiolkovsky became a pioneer of rocketry, an ardent advocate of the conquest of outer space, and the author of his own utopian scheme for the reorganization of the universe.[14]

The physics professor Goddard not only contributed to the scientific basis for liquid-fuel rocketry, he was the first person actually to achieve sustained rocket flight, in 1926. Like Tsiolkovsky, Goddard was also inspired as a child to dream of spaceflight by the writings of Jules Verne, as well as by those of H. G. Wells. (Later in his career Goddard actually wrote a revised version of Verne's *From the Earth to the Moon*, correcting and updating the technical content, and he also tried unsuccessfully to engage the aged Wells in correspondence.) Again like Tsiolkovsky, Goddard was early handicapped by a physical ailment, tuberculosis, which throughout his childhood kept him out of school and out of touch with the world, and which later contributed to his lifelong isolation. According to his biographer, as his health failed, Goddard's youth was "a lonely time"; "his former fellow students were going ahead in school [and] his ailments had left him like the lame boy in the 'Pied Piper of Hamlin,' hobbling to reach the mountain into which his classmates had disappeared. Perhaps he dreamed of devising ways to soar over the mountain. . . ." At the age of seventeen, Goddard had his own epiphany; while sitting high up in a cherry tree, he too thought of using centrifugal force as a means of spaceflight. "As I looked toward the field to the east, I imagined how wonderful it would be to make some device which had even the possibility of ascending to Mars. . . . I was a different boy when I descended the ladder. Life now had a purpose for me."[15]

Both Goddard and Tsiolkovsky "pursued their technological goals with a fervor that can only be regarded as religious," noted the historian of aeronautics Tom Crouch. Tsiolkovsky especially hoped that through his work on space travel he might help foster a new global and cosmic consciousness and thereby bring about "the perfection of human society and its individual members." "Through our technologies, we are subcreators," Walter McDougall wrote in his history of space exploration (aptly entitled *The Heavens and the Earth*). "Hence we have never, from Protagoras to Francis Bacon to Tsiolkovsky, been able to separate our thinking about technology from teleology or eschatology." Before long, the peculiar preoccupation of these single-minded rocket pioneers become the collective obsession of an international cadre of latter-day Gun Club enthusiasts, who, if they aimed also at the stars, set their sights for the most part on more terrestrial targets. Tsiolkovsky's work was put to use by Stalin's military, just as Goddard's was by the U.S. military (Goddard himself, no doubt in pursuit of his loftier aims, eagerly sought military patronage and lent his efforts to military purposes). Meanwhile, parallel scientific work on liquid-fuel rockets by the Transylvanian theoretician Hermann Oberth was put to use with revolutionary effect by the millenarian madmen of the Third Reich, through the inspired wizardry of the young enthusiast Wernher von Braun.[16]

As Johannes Kepler studied at a convent school that was haunted by the spirit of Dr. Faustus, so the young von Braun attended the boarding school of Ettersburg Castle, near Weimar, where Goethe had lived for a time writing his *Faust*. For his confirmation at the age of thirteen, in 1926—the year of Goddard's first successful rocket flight—von Braun received from his parents a telescope, which first turned his attention toward the stars. As he later recalled, he wanted "not just to stare through a telescope at the moon and the planets, but to soar through the heavens and actually explore the mysterious universe. I knew how Columbus had felt."[17]

The following year, von Braun read Oberth's book *The Rocket into Planetary Space* and "fell into a state of supreme elation" about the realistic possibility of actually traveling to the moon and the planets, which became his lifelong obsession. Thus, at the age of fourteen, he wrote a letter to Oberth declaring his dedication to rocket development and space travel, began to build his own rockets, and wrote his first essay about a journey to the moon. Whereas Goddard's youthful exuberance about space travel long remained a solitary preoccupation, von Braun early turned his enthusiasm outward.[18]

At seventeen, he published an article about a manned space station, joined a group of other young experimenters to form a rocket society, and before long became Oberth's assistant. He had begun to realize, however, that indulging his passion for space would be a very expensive proposition indeed, beyond the means of private citizens. Thus, at twenty he opportunistically determined that he had to join the Gun Club, in whose employment he remained for the next thirty years. "It became obvious," von Braun later wrote, "that the funds and facilities of the Army would be the only practical approach to space travel."[19]

Under the terms of the Treaty of Versailles, Germany was not permitted to develop conventional artillery weapons, and so, in 1929, its military decided to pursue rocketry as an alternative approach to long-range warfare. It was into this effort that von Braun was drawn by the army's rocketry chief, Walter Dornberger, becoming a Nazi Party member and ultimately achieving the SS-officer rank of *Sturmbannführer* (major). At von Braun's suggestion, the army's rocket-development center was soon relocated to the remote site of Peenemünde, on the Baltic Sea, where his grandfather had gone duck-hunting. There von Braun served as technical director and perfected the pioneering A series of rockets, which first demonstrated the superiority of rockets over cannons in range, payload, and accuracy. In the fall of 1942, he oversaw the successful test flight of the world's first precision-guided, long-range, rocket-powered missile, the A4, which had been designed to meet the military specifications of a two-hundred-mile range with a metric-ton warhead. After the test, Colonel Dornberger toasted the assembled celebrants: "Ladies and gentlemen, today the spaceship has been born." The Third Reich command, however, had other expectations. Two years later, some three thousand A4 rockets, renamed the V-2 (for "vengeance") by Hitler's propaganda chief Joseph Goebbels and mass-produced with the forced labor of concentration-camp inmates, rained down upon the civilian populations of France, Belgium, and England.[20]

At the war's end, von Braun and his rocket team were taken into protective custody by the U.S. Army, which sought to exploit their expertise for the incipient American rocket-development program. They were transported, along with the captured V-2 rockets, parts, and technical records, to Fort Bliss, Texas, and were there put to work designing and testing rockets for the American military. It was during this sojourn in Texas that von Braun found religion. Raised a Lutheran, he had never been a believer. But here in fundamentalist West Texas, at the prompting of a local Nazarene minister, von Braun began to study the Bible and soon declared himself a born-again Christian, a faith he would continue to profess and proclaim up to his death.[21]

In the wake of the Cold War and the Korean War, the American rocket-development program gained momentum, focusing upon tactical as well as strategic weapons, with both conventional and nuclear payloads. Many of the leading atomic scientists and engineers now devoted their technical talents to delivery systems for their weapons: intermediate and intercontinental ballistic missiles. In 1950, von Braun and his army team were relocated to Huntsville, Alabama, where they eventually became the brains behind the new Army Ballistic Missile Agency at the Redstone Arsenal. There they developed the country's most reliable weapons-delivery systems, the Redstone and Jupiter rockets, and at the same time laid the groundwork for what became the American space program.

As early as the summer of 1954, von Braun had proposed the use of a Redstone rocket to put the world's first satellite into orbit by 1956, but the army's Project Orbiter lost out in the interservice space competition with the Naval Research Laboratory's Vanguard project, and the venture was shelved. (Apparently President Eisenhower did not look favorably upon the Huntsville leadership, nor did he want the satellite effort to be overtly military, for fear of provoking the Soviet Union.) In the panic following the successful Soviet launch of Sputnik in October 1957, however, and repeated Vanguard failures, the army project was revived, resulting in the successful orbiting of the first U.S. satellite, Explorer I, less than three months later.[22]

The earliest history of manned spaceflight followed the same pattern. In August 1957, von Braun had demonstrated the viability of a safe re-entry with a nosecone test using the army's Jupiter C rocket. The following spring, in April 1958, just three months after the launch of Explorer I, the ABMA officially proposed to launch a man into space by the end of 1959. The army project for what would have been the world's first manned spaceflight—a suborbital flight on a ballistic trajectory using a modified Redstone rocket—was called Project Adam. (Also that spring, the ABMA began work on the Saturn booster, which would send men to the moon a decade later.) This time the army project was lost in the shuffle to establish a new civilian space agency; the National Aeronautics and Space Agency (NASA) was created that summer. Before long, NASA had established its own first manned space mission, dubbed Project Mercury. From the start, however, Mercury was modeled upon Adam. NASA quickly worked out an agreement with the army for collaboration with Huntsville and for the use of the Redstone rocket as the Mercury launch vehicle. The first Mercury mission, Alan Shepard's suborbital flight of May 1961, was nearly identical to the one proposed in Project Adam. (Indeed, at Huntsville that mission was still informally referred to as Project Adam, with ABMA's official sanction.) Because of the political delays, however, Mercury's maiden spaceflight came only after that of the world's first—by the Soviet Union's Yuri Gagarin—a month earlier. Thus what was probably the world's earliest—and, despite the political delays, ultimately successful—effort to put a man into space was called Adam.[23]

As the biblical name chosen for this first heavenly ascent attests, religious inspiration, coupled with Cold War competition, fueled the manned-spaceflight effort. Unmanned space vehicles like Explorer, after all, might just as well have provided the scientific and surveillance capabilities sought through manned rockets, and with greater economy and efficiency (in manned efforts, much of the engineering effort and cost was dedicated just to keeping the astronauts alive). Why, then, send men into space? It was God's purpose, wrote von Braun (who named both Adam and Explorer), "to send his Son to the other worlds to bring the gospel to them." Von Braun had come to view spaceflight as a millennial "new beginning" for mankind, the second and final phase of his divinely ordained destiny. The astronaut, the mortal agent of this new "cosmic" era, was thus another Adam, conceived to extend the promise of redemption across the celestial sea.[24]

"Only man," von Braun observed, echoing Kepler, "was burdened with being an image of God cast into the form of an animal," a being at once earthly and heavenly. "And only man has been bestowed with a soul which enables him to cope with the eternal." In 1959—the proposed year of Adam's first flight—von Braun suggested an apocalyptic purpose for mankind's venture into space. "If man is Alpha and Omega, then it is profoundly important for religious reasons that he travel to other worlds, other galaxies; for it may be Man's destiny to assure immortality, not only of his race but even of the life spark itself. . . . By the grace of God, we shall in this century successfully send man through space to the moon and to other planets on the first leg of his last and greatest journey. . . ."[25]

Like his counterparts in the nuclear-weapons business, von Braun had come to view his lethal missiles in millenarian terms. He labored to perfect long-range weapons-delivery systems (his Redstone was the first medium-range nuclear weapon and the first to detonate an atmospheric atomic weapon) and even extolled the virtues of space-based warfare (which "would offer the satellites' builders the most important tactical and strategic advantage in military history"). Nevertheless, at the same time he steadfastly maintained that the ultimate end of mankind's conquest of space was its own salvation. "Here then is space travel's most meaningful mission," he argued shortly after his arrival in Huntsville. "On that future day when our satellite vessels are circling Earth; when men manning an orbital station can view our planet against the star-studded blackness of infinity as but a planet among planets; on that day, I say, fratricidal war will be banished from the star on which we live. . . . humanity will then be prepared to enter the second phase of its long, hitherto only Tellurian history—the cosmic age."[26]

The religious foundation for von Braun's millenarianism was explicitly Christian. "When man, about 2000 years ago, was given the opportunity to know Jesus Christ, to know God who had decided to live for a while as man amongst fellow men, on this little planet," von Braun later wrote, "our world was turned upside down through the widespread witness of those who heard and understood Him. The same thing can happen again today."[27]

Von Braun's religious convictions were confirmed, not contradicted, by his scientific and technological undertakings. Throughout his American career, he adamantly insisted that science and technology were compatible with, and essential to, the achievement of religious ends. "In this reaching of the new millennium through faith in the words of Jesus Christ, science can be a valuable tool rather than an impediment," he maintained. "It has frequently been stated that scientific enlightenment and religious beliefs are incompatible," he said in a commencement speech in 1958; "I consider it one of the greatest tragedies of our times that this equally stupid and dangerous error is so widely believed." "Science and religion are not antagonists," he later argued. "On the contrary they are sisters. While science tries to learn more about the Creation, religion tries to better understand the Creator. Speaking for myself, I can only say that the grandeur of the cosmos serves only to confirm my belief in the certainty of a Creator." "Today, I am a Christian," he wrote to a correspondent. "Understanding the nature of creation provides a substantive basis for the faith by which we attempt to know the nature of the Creator. My experience with science, then, led me to God—it was as if I was putting a face on God."[28]

Like so many of his faithful scientific forebears, von Braun held to a firm belief in immortality—"the continuity of our spiritual existence after death"—which was grounded upon the precedent of Christ's resurrection. "In our search to know God," he explained, "I have come to believe that the life of Jesus Christ should be the focus of our efforts and inspiration. The reality of this life and His resurrection is the hope of mankind." In this spirit, he supposed that "a human being is so much more than a physical body that withers and vanishes after it has been around for a number of years. It is inconceivable to me that there should not be something else for us after we have finished our earthly voyage."[29]

It can be claimed that Wernher von Braun was an opportunist who delivered death, if need be, in the name of, and in determined pursuit of, extraterrestrial transcendence. Thus, he became a rocket warrior for the Third Reich and, in the process, a Nazi Party member and SS officer. Thus, he later was the architect of the U.S. Army's long-range ballistic-missile arsenal and, in the process, a patriotic born-again Christian. Nevertheless, and even though his religious pronouncements sometimes seem a bit prepackaged, it is quite likely that his belated beliefs became genuine.

"The significance of religious thinking dawned on me rather late," he told a newspaper reporter in 1968. "I started reading religious books and the truth of Christ's teaching emerged like a revelation." At Huntsville, he joined the Episcopal Church of the Nativity, enrolled his daughters in Bible study, and wrote and spoke publicly about his religious convictions, especially to youth groups. His closest friends and neighbors attested to his religious sincerity. In Huntsville, he attended prayer breakfasts at the Redstone Arsenal and the Marshall Space Flight Center, which had been initiated by the International Christian Leadership, and gave a keynote address at that organization's thirtieth-anniversary convention in 1965.[30]

At least some of von Braun's scientific colleagues had misgivings about his religious inclinations, precisely because of his sincerity. "It was surprising to some of von Braun's associates that in spiritual matters, he would reach so deeply into the realm of the irrational," physicist Ernst Stuhlinger, one of von Braun's closest scientific collaborators at both Peenemünde and Huntsville, recalled. "His entire work for space was built upon the exact laws of the natural sciences. . . . In his religious beliefs, it was different. He did not enter into discussions of the points he made. . . . 'Matters of faith are not really accessible to our rational thinking,' he would say, 'I find it best not to ask any questions, but simply to believe. . . . It is best not to think. . . .'" Despite the doubts of some scientists and engineers, however, von Braun was by no means alone in his religiosity. Indeed, among the space community at Huntsville and elsewhere, his beliefs were widely shared, and he was more the norm than the exception.[31]

Perhaps most important, von Braun enjoyed the full support of his commanding officer at the ABMA, General John B. Medaris, who was himself a devout Christian. Medaris is sometimes called the true father of the U.S. space program because of his role in overseeing the pioneering development of the Redstone and Jupiter rockets, the first U.S. satellite (as well as Project Adam, which he justified as a means of troop transport), and the Saturn booster. In 1960, Medaris resigned from the army in frustration over interservice space rivalries and the establishment of NASA, which obstructed and finally put an end to the ABMA's role in space. After a stint in private industry, he became first a lay reader and deacon and then an ordained minister of the Episcopal Church of the Good Shepherd, near Cape Canaveral, Florida, and later an Anglican Catholic priest. "I think it was in England in 1942 or 1943 that I became convinced of the power of the Lord," Medaris recounted. The responsibilities of his postwar commands, as well as repeated trials with cancer beginning in 1956, heightened his religious convictions. "No one could have had the continuing success in the space area that I did without God's help." In 1957, at the time of the Project Adam proposal, Medaris wrote a prize-winning essay entitled "A General Looks at God." In the early 1970s, he headed an ultimately unsuccessful effort to establish a Chapel of the Astronauts adjacent to the Kennedy Space Center.[32]

The religious environment of the space community at Huntsville was no doubt encouraged by such outspokenly devout leadership, as well as by the generally religious population of northern Alabama. In addition to the annual mayor's prayer breakfasts, the first of which was held at the Redstone Arsenal, prayer groups routinely met in the mornings, before the workday, at both the Redstone Arsenal and the Marshall Space Flight Center. In 1969, Billy Graham held a rally at the Redstone Arsenal. After the departure of von Braun as Marshall's director in 1970, this religious ethos was sustained by his successors, notably William R. Lucas, who became the director of the center in 1974. A metallurgist and veteran of the von Braun team since 1952, Lucas did pioneering work on the heat shield used for rocket-nosecone re-entry, and later served the Huntsville space program in various high-ranking scientific and administrative capacities. He was also a lay reader active in the Baptist Church and an articulate advocate of the integration of science and religion, about which he gave speeches at evangelical gatherings, including the Billy Graham crusade. As the "crown of God's creation," mankind was expected to learn all he could about creation, Lucas argued, and space exploration was one way to go about it. Like von Braun, Lucas also saw space technology as a means of spreading the good news, and suggested early on that rockets and satellites could be used with unprecedented effectiveness to broadcast the gospel.[33]

"I didn't feel unusual in this community," Lucas recalled. "The vast majority of people at Marshall, and before that at the ABMA and the Redstone Arsenal, were Christian people." In the space community at Huntsville, "the oddity was not the believer but the nonbeliever." Huntsville NASA scientist Rodney W. Johnson, a planner for lunar missions, who likewise sought to bring science and religion more closely together, concurred. "My contacts indicate that a surprising number of scientists, engineers, and technicians associated with the space program have a deep and vital faith. More, proportionately, than in any other fields and professions." Johnson himself viewed the flights to the moon as a "reminder that man is made in God's image" and that the heavens are not just God's domain, but mankind's as well.[34]

This same religious ethos of the space community clearly manifested itself at Cape Canaveral in the early 1970s, when General Medaris's effort to establish the Chapel of the Astronauts garnered strong support not only from the robust local evangelical community but also from many within the Kennedy Space Center, NASA's prime launch site, including its director Kurt Debus, a veteran of the original von Braun team. After his retirement from the Kennedy Space Center, electrical engineer Edwin Whisenant, who had also been involved in the early rocket launches as well as the moon-landing program, devoted himself to the analysis of biblical prophecy. In the 1980s, he wrote several books predicting (unsuccessfully) the time of the Rapture. "It's an obsession," he said. "The time is short. I'm telling people the end is near and to get their children and everyone they care about under the blood of Jesus."[35]

The same spirit was amply evident at the Johnson Space Center, near Houston, the mission-control center for manned spaceflights and home base and training center of the astronauts themselves. Bible-study groups proliferated throughout Johnson, in the simulation and training departments, the astronauts' office, and Mission Control itself. "There are a lot of Christians at Johnson," noted Jerry Klumas, a veteran systems engineer and cofounder of the nearby NASA Church of the Nazarene. "The Christian community at NASA is not a minority; it is very significant, and NASA people are outspoken about being Christians."[36]

In Klumas's view, following the prophecy of Daniel, the great increase in knowledge generated by space exploration signals that the end-times are at hand. Moreover, he observed, as the speed of space travel accelerates, aging decreases, and the traveler nears immortality. "The spiritual laws governing our salvation have always been in existence, but I had to discover them for myself and learn how to put them into action," declared lunar-landings project engineer Robert Bobola. "How can a man trained in the sciences believe in God? According to the evidence, I have to believe in Him, I've checked Him out personally, and He's for real."[37]

NASA aerospace engineer Tom Henderson was involved from the beginning in all of the manned space programs, designing mission simulators and training the astronauts. At the same time, for twenty-five years he has been an active evangelical Christian, preaching the gospel of creationism throughout the hemisphere. Many of his colleagues are also creationists, including Maury Minette, who helped to train Neil Armstrong and Edwin Aldrin in their mission simulations for the first landing on the moon. "To me, science as a whole points to God," declared Tom Henderson. And it also contributes to a recovery of mankind's lost knowledge. "I think Adam was brilliant," Henderson noted, but the preflood civilization he started was lost, and "mankind has had to climb the hill of knowledge" once again. "When Christ returns," however, "to rule for a thousand years, the earth will return to its preflood state. . . . Either when I die or when the Rapture of the Church occurs, whichever happens first, I will return to earth with Christ; with a new immortal body I will live on earth but not as a man; I will be able to travel in space without a spaceship; I will meet with Robert Boyle and Isaac Newton."[38]

According to Jerry Klumas, expression of religious beliefs was quite acceptable at NASA. "NASA administrators do not discourage such behavior. NASA is not hung up about separation of church and state. At Johnson, administrators encourage Bible-study groups, providing them with meeting rooms. Just about every leader of NASA is an active church member." This official sanction of religious practice at Huntsville, Houston, and Cape Canaveral mirrored the sentiment at NASA headquarters in Washington.[39]

Hugh Dryden, the first operational chief of NASA in its formative years, was a licensed Methodist lay preacher as well as an esteemed scientist, and, like so many others at NASA, he maintained that there was no necessary conflict between the two identities. A brilliant aerodynamicist, Dryden was a central figure in both the establishment of NASA and, in particular, the push for manned spaceflight. He served for a decade as director of the National Advisory Committee on Aeronautics before becoming NASA's first deputy administrator in 1958, a position he held until 1965. Throughout his life, he was an active member of the Calvary Methodist Church, where he regularly gave sermons and taught Bible-study classes. In 1962, he was named "Methodist layman of the year."[40]

Dryden's sermons resounded with the transcendent strains of the religion of technology. "Of all the expeditions of the human mind and soul into the great mysteries of life," Dryden preached, "I know of none so bold as the search of man to find God." One of his favorite themes, which he repeated in his sermons, was mankind's "birthright, creation in the image of God." We are "made in the image of God, a little lower than the angels," he insisted. It was this endowment which gave men "the ability to rise above life on a purely physical plane to the realm of the mind, and to increase his intellectual powers, his power to think, to comprehend, and to reason." "God has shared with us some of his creative power," Dryden declared, including the powers of science and technology. "By all means seek Him in nature. The more we understand of nature the more we comprehend the intellectual state of its Creator."[41]

Much of this divine gift had been lost through sin, Dryden noted, "but like the old masterpiece of the painter, the original image can be restored. By patient, careful effort, we may, if we will, begin to bring out those elements in our character which are God-like. To this task we are challenged by the life of Jesus Christ, who demonstrated to us what we might hope to become." "Would that our leaders today and the rest of us who follow could discover and understand clearly our tasks, and pursue them with the aim that our hands, our lips, our brains might be the channels through which the Kingdom of God may come."[42]

Equally fervent in his religious convictions was the only two-term NASA administrator, James Fletcher, a devout member of the Church of the Latter-Day Saints (Mormons). A physicist by training, Fletcher devoted much of his scientific career to the development of long-range weapons-delivery systems, under the auspices of both the Department of Defense and private industry. At NASA, he was "generally recognized as one of the most influential administrators from the first three decades of space flight." According to NASA's own chief historian, "Fletcher's approach toward directing the U.S. space program owed something to his Western American and Mormon conceptions of the world. This heritage came into play throughout Fletcher's NASA career as an underlying philosophy of why humans should explore space," an endeavor he described as a "God-given desire." His Mormon beliefs led him to envision space exploration as "an intellectual frontier of expanding knowledge and the progress of understanding about nature and, by extension, about divinity." Because of his Mormon belief in the existence of a plurality of worlds, Fletcher strongly promoted space programs aimed at the search for extraterrestrial intelligence, such as the Viking mission to Mars and the SETI (Search for Extraterrestrial Intelligence) Program. (In the same spirit, Bruce Murray, director of NASA's Jet Propulsion Laboratory, declared in 1979 that "the search for extra-terrestrial intelligence is like looking for God.") More important, Fletcher's strongly religious orientation led him to lend full headquarters support to the various religious currents within NASA. By the time of Fletcher's appointment in 1971, public controversy about religion in NASA had rendered such official support quite explicit.[43]

On Christmas Eve 1968, the astronauts on Apollo 8—the first manned mission to the moon—broadcast back to earth their reading of the first ten lines of the Book of Genesis. Three days later, Madalyn Murray O'Hair, the militant atheist whose lawsuit had resulted in the 1963 Supreme Court ban on required prayers in the public schools, vigorously protested such religious display on the part of a governmental scientific agency. "It's incredible," she exclaimed, "men who are supposed to be scientists reading from Genesis like that." Seven months later, in August 1969, she formally filed suit against NASA, seeking an injunction against its "permitting religious activities or ceremonies," which she decried as "an attempt to establish the Christian religion of the U.S. government before the world."[44]

The defendant in the suit was Fletcher's predecessor as NASA administrator, Thomas O. Paine, an Episcopalian. NASA's official legal position was that the astronauts were merely exercising their own religious rights and that NASA would neither direct nor restrict any such activities. Speaking before the National Press Club on the day the suit was filed, however, Paine went a step further in his support of the astronauts' actions. "The fact that on Christmas Eve Frank Borman and his crew read aloud the opening lines of Genesis . . . undoubtedly gave some offense to Mrs. O'Hair," Paine noted. "But to my mind, it was a proper and fitting thing to do." Behind the scenes, Paine's administration encouraged a show of public support for the religious reading, and soon received over a million citizen petitions from a religious radio network.[45]

O'Hair's suit was dismissed by the Federal District Court in December 1969; her first appeal was denied by the Fifth U.S. Circuit Court of Appeals seven months later, and the U.S. Supreme Court declined to hear her last appeal in March 1971. (In March 1973, she filed another suit to prevent prayer services in Congress and the White House, which was also dismissed.) Although unsuccessful, the legal challenges did make Congress (and thus NASA) more cautious with regard to overt support of religious causes. It was no doubt for this reason that General Medaris's effort to construct a Chapel of the Astronauts (dedicated to "worship of the Creator and Praise of the Almighty") on public land adjacent to the Kennedy Space Center, which required congressional action, finally had to be abandoned, despite strong support from many members of Congress, the local community, and such leading religious figures as Billy Graham. Both Paine and Fletcher had strongly endorsed the project.[46]

If NASA had to become somewhat more cautious in public, official expression of overtly religious sentiment continued in private, especially during Fletcher's regime. For several years following the O'Hair litigation, NASA's Office of Public Affairs received and responded to many letters from private citizens regarding the religious controversy. Most of the responses were quite general, and only hinted at the religious sentiments of NASA officialdom: "We thank you for your interest and know you will be relieved to know that the astronauts are now legally as well as spiritually free to express themselves." In June 1992, however, the director of the Office of Public Affairs, O. B. Lloyd, NASA's official spokesman, became far more explicit. A woman had written NASA to express her concern about the lack of "spiritual thought" in Apollo 16. Lloyd reminded her of the Apollo 8 Genesis reading and quoted from Psalm 8, the prayer that had been recited by Edwin Aldrin on Apollo 11, the first lunar-landing mission. He also referred to the recent decision by Apollo astronaut James Irwin to establish his own evangelical ministry. "These certainly demonstrate the spiritual emphasis brought to the space program by the astronauts," Lloyd wrote. "We agree with you," he wrote on behalf of NASA, "and I know the astronauts do too, that the Apollo missions could not have succeeded without the help of God. . . . I believe that you can be reassured that those who work in the space program are indeed aware of the presence of the Creator and are not neglectful of spiritual values."[47]

In February 1974, a stained-glass "space window" was officially installed at the Washington Cathedral, containing a two-inch-diameter lunar rock sample brought back on Apollo 11. The window had been paid for by now private citizen Thomas Paine. At the dedication ceremony on the fifth anniversary of the first lunar landing, Paine read one of the lessons while Fletcher read another. The *NASA Headquarters Weekly Bulletin* announced the event, pointing out that the dean of the cathedral "will preach on the spiritual significance and the religious implications . . . of the first journey from the planet Earth." "Should we hesitate to exploit the first step?" George Mueller, director of NASA's manned spaceflight program, had asked after the first landing on the moon, giving voice to the apocalyptic and millenarian spirit that infused the whole enterprise. "Should we withdraw in fear from the next step, should we substitute temporary material welfare for spiritual adventure . . . ? Then will Man fall back from his destiny, the mighty surge of his achievement will be lost, and the confines of this planet will destroy him."[48]

On numerous occasions during the first decades of the space program, NASA scientists and engineers gave expression to their religious beliefs, with official sanction if not financial support. In 1958, after several failures with their rocket, the designers of the Vanguard finally wired a St. Christopher medal to the base of a gyroscope package in the second-stage guidance system. The design modification was dutifully described in detail on the required specification form, which was officially signed off by the necessary personnel, with the specified objective: "Addition of Divine Guidance." In 1973, Josef Blumrich of the Program Development Directorate at the Marshall Space Flight Center patented a design for an omnidirectional wheel inspired by a description by the Old Testament prophet Ezekiel. In 1974, NASA's

Earth Observations Program became involved in an effort to use satellite images to locate Noah's Ark on the top of Mount Ararat in Turkey; and in 1979, NASA technicians at the Jet Propulsion Laboratory used their Viking and Voyager equipment and expertise to test the Shroud of Turin—allegedly Christ's burial cloth—for authenticity. (NASA Headquarters was careful to point out that "probing the mystery of how the shroud got its image is not a government-sponsored project"—the tests were not paid for by NASA.)[49]

If NASA ground technicians and administrators sometimes used their expertise and authority to give expression to their beliefs, the religious significance of the space program was given greatest expression through the words, actions, and personas of the ascending saints themselves. "Only a mixture of adventurous impulses and religious convictions of the deepest sort would persuade normal warm-hearted human beings, such as many astronauts seem to be, to take part in such a life-denying ritual," Lewis Mumford observed. "Besides high physical courage and the promise of an early termination of the ordeal, they need a deep religious conviction, all the more serviceable if unconscious of their role as Heavenly Messengers."[50]

The first American astronauts were all devout Protestants, a fact that was officially advertised in the early days of the manned spaceflight program to distinguish America's pious effort from that of its Soviet rival, with its creed of "godless communism." Indeed, as Soviet cosmonauts triumphantly declared that they had not encountered God, some enthusiasts for the American space program insisted that nonbelievers had no place in it. At a military chaplains' convention in 1963, Brigadier General Robert Campbell declared that "there is no room for agnostics in America's space and missile program." "If we must make our missiles work with agnostics, then we should join the other side," argued Air Force Colonel Sam Bays. Through the end of the Apollo and Skylab programs, 90 percent of the men chosen to be astronauts had been active Christians, and of these 85 percent belonged to Protestant denominations.[51]

"I of course am a Christian," declared the pilot of the renamed Project Adam, Alan Shepard, who attended the Christian Science Church. I "take my religion very seriously," said John Glenn, the first

American to go into earth orbit. An active Presbyterian who taught Sunday school, Glenn assured Congress upon his return of his steadfast "devotion to God." He explained that the observed "orderliness" of the universe, the existence of "a definite plan," "shows me there is a God," and that "He'll be wherever we go." "It wasn't just an accident. And, although we can't weigh and measure God in scientific terms, we can feel and know Him. More important, we can let Christian principles guide our lives. When we do this—when we believe in God and the teachings of Christ—we see the results."[52]

"I am a Christian, a Methodist," Gordon Cooper told Congress. "I named my spacecraft Faith 7, first, because I believe in God." During his earth orbit, Cooper became the first astronaut to recite a prayer in space. "I would like to take this time to say a little prayer for all the people, including myself, involved in this launch and this operation. . . . Father, help guide and direct all of us that we may shape our lives to be much better Christians." On his Mercury mission, Cooper carried with him in his flight suit a hand-made Christian flag in white silk with a red cross on a blue field. "I consider myself religious," said Virgil Grissom; "I am a Protestant, I belong to the Church of Christ, I consider myself a good Christian." Scott Carpenter described himself as a man of "religious faith" and a "faithful church-goer." Walter Schirra was active in the Episcopal Church; Deke Slayton, later head of the Astronauts' Office in Houston, was Lutheran.[53]

The spiritual convictions of the first generation of astronauts were more than matched by those of their successors in the Apollo, Skylab, and Shuttle programs, who frequently gave voice to the religion of technology. The Christmas 1968 voyage of Apollo 8, the first time astronauts left earth's vicinity to orbit the moon, was declared a "millennial event" by Pope Paul VI, and the astronauts themselves echoed the message. Their Christmas Eve reading from Genesis was not spontaneous but rather was written into the flight plan by Frank Borman, a lay reader of St. Christopher's Episcopal Church in Seabrook, Texas. The moon voyage was "the final leg in my own religious experience," Borman recounted; "I saw evidence that God lives." On the third lunar orbit, he radioed a prayer from space, which he dedicated "to Ron Rose and the people at St. Christopher's." His crewmate James Lovell had been converted to the Episcopal Church by Borman. Of his lunar experience he later declared, "I can't think of a better religious aspect of the flight than to further explore the heavens."[54]

"You think about what you're experiencing and why . . . are you separated out to be touched by God . . . ?" reflected Apollo 9's Rusty Schweikert, who later turned to Transcendental Meditation, Zen Buddhism, and New Age cosmic consciousness, and lent his support to the apocalyptic Biosphere II effort to create a new artificial habitat for humanity for use on another planet after the demise of the moribund earth. The Apollo 10 astronauts brought along their own Bible; reflecting about his first lunar experience (he later became the last man on the moon), Gene Cernan, a Roman Catholic, confirmed his conviction that "there was no question there had to be some creator of the universe."[55]

Although Neil Armstrong, the commander of the first lunar-landing mission (Apollo 11) had been raised in the Evangelical Church, he did not consider himself especially religious, unlike his fellow crew member Edwin Aldrin, who was a Presbyterian elder and Sunday-school teacher. Before the flight Aldrin took communion with his pastor, who told him, as Aldrin later recalled, that he and his colleagues would "view the earth from a physically transcendent stance," and through their effort "mankind would be awakened once again to the mythic dimensions of man." As the landing module sat in the Sea of Tranquillity shortly after the lunar landing, before Armstrong and Aldrin ventured out onto the lunar surface, Aldrin asked Mission Control for radio silence. He then proceeded to unwrap a small kit provided by his pastor which contained a vial of wine, some wafers, and a chalice, and took communion, reading from John 15:5. "It was interesting to think," he later observed, "that the very first liquid ever poured on the moon and the first food eaten there were communion elements." Later, with the radio on, he read from Psalm 8.[56]

In the orbiting command ship *Columbia,* Michael Collins had reveries of his own. Collins, an Episcopalian, observed that the floor plan of *Columbia* appeared to him like a "miniature cathedral," reminding him of the National Cathedral, where he had been an altar boy. "Certainly, it is a cruciform, with the tunnel up above where the bell tower would be, and the navigation instruments at the altar. The main instrument panels span the north and south transepts, while the nave is where the center couch used to be." After the return of Apollo 11 from the moon, President Richard Nixon proclaimed: "This is the greatest week since the beginning of the world, the Creation." (He was later reminded by his personal religious adviser, Billy Graham, that there had been three greater events than this—Christ's birth, crucifixion, and resurrection.)[57]

On Apollo 12, Pete Conrad took to the moon a Christian flag emblazoned with a cross. Alan Bean carried a Bible as well as a banner provided by the Clear Lake Methodist Church at his request, embroidered with symbols of the Trinity, a Luther rose, the crusader's cross, a chalice, and a Bible. Upon his return, he said that the experience had confirmed his faith in the existence of God. The crew of the ill-fated Apollo 13, which never made it to the moon because of the explosion of an oxygen tank, took with them hundreds of Bibles on microfilm on behalf of the Apollo Prayer League of Houston, which had hoped later to distribute them among the faithful. Apollo 14 Commander Edgar Mitchell, who conducted a telepathy experiment from the moon (he later established the Noetics Institute to pursue research in psychic phenomena), had a Bible in his space suit which he left on the lunar surface along with microfilm containing the first verse of Genesis in sixteen languages.[58]

The moonwalkers of Apollo 15 were among the most religious-minded of the Apollo astronauts. Commander Dave Scott, who drove the lunar rover miles across the barren moonscape, upon his departure left a small red Bible on the top of the rover's control panel. Scott's fellow traveler Jim Irwin, meanwhile, who recited the first verse of Psalm 121 while wandering amid the mountains of the moon, felt a great "closeness to God" and even imagined himself "looking at the earth with the eyes of God." "On the moon the total picture of the power of God and His son, Jesus Christ, became abundantly clear to me. . . . Apollo 15 explored the surface of the moon

with the power of God and Jesus Christ," he declared later. Return-
ing from the moon with the so-called Genesis rock—a four-and-a-
half-billion-year-old lunar sample—Irwin brought back as well a new
appreciation of "the rock of the Word of God." "Jesus Christ walking on the earth is more important than man walking on the moon,"
he insisted. A longtime born-again Methodist turned Baptist, Irwin
became a Baptist minister and created his own evangelical ministry,
which he called High Flight. "I established High Flight in order to tell
all men everywhere that God is alive, not only on earth but also on
the moon," he later explained. Tirelessly speaking and writing on behalf of the evangelical cause (including for Billy Graham's crusades),
traveling to the Holy Land, and leading six expeditions to Mount
Ararat in search of Noah's Ark, Irwin exemplified what ex-astronaut
Brian O'Leary described as the "astronauts' Messiah complex."[59]

Charlie Duke, who as "Capcom" had guided the first lunar
landing from Mission Control, went to the moon himself on Apollo
16 carrying a prayer that he later gave to his Episcopal church.
Duke eventually became a born-again Christian fundamentalist, creationist, and evangelist, and president of the Duke Ministry for
Christ. "That walk on the moon lasted three days, but my walk
with Jesus will last forever," he declared. (Duke's Apollo 16 crewmate John Young shared his religious convictions.) After his return
from the final lunar mission, Gene Cernan reported confidently that
"seeing what I saw . . . I know there has to be a Creator of the universe. . . . It is too beautiful to have happened just by accident."[60]

After the Apollo moon missions, many NASA Skylab and Shuttle astronauts continued to bring to space exploration an abiding
religious faith. Jack Lousma, Skylab and Shuttle veteran, was a
"deeply religious man," a strong evangelical Christian throughout his
life. "Even the Columbia's [space shuttle] guidance system neatly
illustrates how God steers a Christian," he told a reporter for the
Christian magazine *Guideposts*. "God has a reference trajectory for
each of our lives." Don Lind, one of the oldest astronauts to fly, was,
like James Fletcher, a Mormon who devoted much of his time to
evangelical missionary work. Skylab astronaut Donald Pogue later
joined Irwin's High Flight evangelical ministry.[61]

"We must accept the fact that our very existence and where we live is because of God's blessings and his creation," declared Shuttle astronaut Bill Nelson, who later became a Florida congressman. Dave Leestma, who flew on three Shuttle missions, was another evangelical Christian, whose view of the earth from space gave him "clear evidence of creation." Joe Tanner, who taught Leestma's son in Bible school, attributed his selection as a Shuttle astronaut to God's plan and also viewed his experiences in space as confirmation of creation and the infinite nature of God. "I knew that God's hand had always been directing me," agreed Shuttle astronaut Robert Springer, a devout member of the Calvary Bible Church.[62]

Following the examples from Apollo 11, Tom Jones and his crew shared communion in the Shuttle spacecraft, which Jones described as "the most magnificent cathedral you can go to church in." "Being in space was a real religious experience for me," said Jones. "I think there is a Creator, and he did a great job on our planet." Awed by the seemingly superhuman achievements of the space program, Jones considered the success of the enterprise to be as much the work of God as of man. "Clearly a larger hand was at work," he insisted; "it was divinely inspired." Such was the belief as well of Johannes Kepler, who nearly four centuries earlier had first envisioned the prospect of such heavenly ascent. "Should the kind of Creator who brought forth nature out of nothing," Kepler wondered, "deprive the spirit of man, the master of Creation and the Lord's own image, of every heavenly delight?" Apparently not.[63]

CHAPTER 10

THE IMMORTAL MIND:

ARTIFICIAL INTELLIGENCE

If space travel released some men from gravity's grasp and the confines of their earthbound "exile," the divine spirit of mankind nevertheless remained moored by its own bodily incarnation, from which only death could provide deliverance. "Once I measured the skies; Now I measure the earth's shadow. Of heavenly birth was the measuring mind; In the shadow remains only the body," Kepler wrote in verse for his own epitaph. For "the measuring mind," that immortal vestige of mankind's image-likeness to God, true "heavenly delight," Kepler believed, had to await the end of its embodied existence. Just ten years after Kepler's famous dream of a lunar voyage, however, a different dreamer imagined the possibility of such deliverance without death, by means of a deliberate intellectual effort to purify the mind and purge it of its corporeal impediments. Like that of Kepler, René Descartes's dream long inspired much reflection and anticipation but had to wait three centuries for its fulfillment.[1]

Like Kepler, Descartes perceived the mind as mankind's heavenly endowment and, in its essence, distinct from the body, the burden of mortality. "The first thing one can know with certainty," Descartes wrote in a letter, is that man "is a being or substance which is not at all corporeal, whose nature is solely to think." For Descartes, the human intellect was godly—"doubtless received from God," Descartes declared—and was defined by precisely those characteristics which the human being shared with God. "That all things that we very clearly and very distinctly conceive of are true, is certain only because God is or exists and that He is a perfect Being, and that all that is in us issues from Him," Descartes wrote. "If we did not know that all that is in us of reality and truth proceeds from a perfect and infinite Being, however clear and distinct were our ideas, we should not have any reason to assure ourselves that they had the perfection of being true."[2]

The body, on the other hand, reflected mankind's "epistemological fallenness" rather than its divinity, and stands "opposed to reason." Impediments to pure thought, the body's senses and passions deceive and disturb the intellect. "The body is always a hindrance to the mind in its thinking," Descartes argued, which is "contradicted by the many preconceptions of our senses." In the wake of Copernicus and Galileo, Descartes was keenly aware that mere sense-perception could not provide a true scientific understanding of the universe and might indeed retard such understanding. Likewise, the passions ignited by the Reformation had distorted discourse beyond reason and generated confusion and doubt about reliable sources of religious authority and conviction.[3]

In search of some certainty, Descartes sought refuge in pure thought. Though philosophers had long lamented the liabilities which the body posed for the mind, none before Descartes had actually defined the two as radically distinct and mutually exclusive. In so doing, he aimed to emancipate the divine part of man from its mortal trappings, the "prison of the body," and the commotion of the "animal spirits." The human mind at birth "has in itself the ideas of God, and all such truths as are called self-evident," Descartes argued. "If it were taken out of the prison of the body, it would find [these ideas] within itself." He thus proposed a new regime for the intellect, a set of "rules for the mind," designed to cleanse it of bodily impurity and make way for the clear and distinct ideas which humans shared with

God. (Like many of his contemporaries, such as Bacon, Comenius, Wilkins, and Glanvill, Descartes dreamed also of a universal language based upon such precise concepts—a restoration of the prelapsarian, pre-Babel language of Adam—which would help overcome the confusion and conflict engendered by miscommunication.)[4]

"Even those who have the feeblest souls can acquire a very absolute dominion over all their passions if sufficient industry is applied in training and guiding them," Descartes insisted, in monklike fashion. A clear and distinct understanding of the mind's primary notions "cannot be perfectly apprehended except by those who give strenuous attention and study to them, and withdraw their minds as far as possible from matters corporeal. . . . I shall now close my eyes, I shall stop my ears, I shall call away my senses, I shall efface even from my thoughts all images of corporeal things." Only through such training and discipline, Descartes maintained, could a person learn to "think without the body," and thereby achieve "pure intellection," "pure understanding." Descartes viewed geometry and arithmetic—products of the "measuring mind" of the mathematician—as the models of such pure thought, because they "alone deal with an object so pure and uncomplicated, that they need make no assumptions at all which experience renders uncertain." Of "heavenly birth," they lie, as it were, beyond experience, and hence closer to God.[5]

Descartes's peculiar obsession became the principal philosophical preoccupation for three centuries, as diverse thinkers sought to comprehend the mechanisms of human understanding, the categories of reason, the phenomenology of mind. Moreover, in the mid-nineteenth century, mathematics became not just a model for pure thinking but the means of describing the process of thought itself. In 1833, at the age of seventeen, the mathematician George Boole had what he described as a "mystical" experience. "The thought flashed upon him suddenly one afternoon as he was walking across a field [that] his main ambition in life was to explain the logic of human thought and to delve analytically into the spiritual aspects of man's nature [through] the expression of logical relations in symbolic or algebraic form."[6]

An intensely religious man (like Newton, an Anglican with Unitarian tendencies), Boole had originally intended to join the clergy, but the death of his father compelled him instead to seek employment as a teacher. Like Descartes, Boole believed that human thought was mankind's link with the divine and that a mathematical description of human mental processes was therefore at the same time a revelation of the mind of God. "We are not to regard Truth as the mere creature of the human intellect," he argued. "The great results of science, and the primal truths of religion and of morals, have an existence quite independent of our faculties and of our recognition. . . . It is given to us to discover Truth—we are permitted to comprehend it; but its sole origin is in the will or the character of the Creator, and this is the real connecting link between science and religion. It has seemed to be necessary to state this principle clearly and fully, because the distinction of our knowledge into Divine and Human has prejudiced many minds with the belief that there is a mutual hostility between the two—a belief as injurious as it is irrational." The purpose of his study of mathematics and nature, Boole insisted, quoting Milton, was simply "to justify the ways of God to Man."[7]

According to his biographer, "It is impossible to separate Boole's religious beliefs from his mathematics." His binary algebra, in which the number one symbolized the universal class, quite possibly reflected his Unitarian belief in the unity of God and the oneness of the universe. It was this algebra which Boole developed to describe the mathematical foundation of human thought (and which later became the logical foundation of digital computers). In his seminal work *An Investigation of the Laws of Thought, On Which Are Founded the Mathematical Theories of Logic and Probabilities* (1854), Boole declared "the truth that the ultimate laws of thought are mathematical in their form." Even in this highly technical treatise, Boole's belief in "another order of things" and his reverential view of human thought as a reflection of the divine are evident.[8]

"The progress of natural knowledge tends toward the recognition of some central Unity in Nature," wrote Boole, "a primal unity." And "human nature, quite independently of its observed or manifested tendencies, is seen to be *constituted* in a certain relation to the Truth; and this relation, considered as a subject of speculative knowledge, is as capable of being studied in its details, is, moreover, as worthy of being so studied, as are the several departments of physical science, considered in the same aspect." (Boole was a great admirer of Newton, who certainly viewed his own scientific efforts in this light.) "We cannot embrace this view without accepting at least as probable the intimations which . . . it seems to furnish respecting another and a higher aspect of our nature."[9]

What was barely suggested in the guarded language of the scientist was more fully expressed in verse. During the same years in which he wrote his *Laws of Thought,* Boole composed several poems which reflected his celestial view of human thought. "Space diverse, systems manifold to see, /Revealed by *thought* alone; was it that we, /In whose mysterious spirits thus are blent, /Finite of sense and Infinite of *thought,*" he wrote in his "To the Number Three." Just before the publication of *Laws of Thought,* he wrote "The Communion of the Saints," a tribute to his scientific forebears ("an inseparable band" in "spirit-land," he described them in another poem).

> Then the dead, in *thought* arriving
> From far-off regions bright,
> Seem to aid our earnest striving
> For the holy and the right. . . .
>
> Seeker after Truth's deep fountain,
> Delver in the soul's deep mine,
> Toiler up the rugged mountain
> To the upper Light Divine,
> Think, beyond the stars there be
> Who have toiled and wrought like thee.[10]

Descartes had strived to divorce the mind from the body in order to insulate thought from corporeal distortion and make possible the formulation of clear and distinct ideas, the foundation of true knowledge. He believed that his philosophical method might help mankind overcome the epistemological handicaps of its fallen state and regain control of some of its innate godly powers. Boole's inspired effort to represent the human thought process in mathematical terms pushed this perfectionist project further. Now precise logical analysis could serve as an aid to the mental discipline Descartes demanded, providing a new set of "rules for the mind," for clarifying ideas. Within half a century, mathematical logicians such as Gottlob Frege, Bertrand Russell, and Alfred North Whitehead had greatly improved upon Boole's effort, laying the basis for a mathematical calculus of human reason.

At the same time, the reduction of human thought to mathematical representation made imaginable the mechanical simulation or replication of the human thought process. For, once the mysteries of the immortal mind were rendered transparent and comprehensible, they might be mechanically reproduced, and thereafter independently manipulated. The thinking person might then be joined by the thinking machine modeled upon the patterns of human thought but independent of the thinking person. The "measuring mind" of man could take form in a new, ultimately more durable medium. What Descartes called "thinking without the body" would now take on potent new meaning.

The inspiration behind this peculiar project remained religious, even after the explicit vocabulary and professions of faith had given way to technical jargon. A thinking machine that replicated the defining characteristic of the human species, *Homo sapiens*, would not, as many supposed, represent an irreverent deprecation of humanity in favor of mechanism, nor would it constitute a celebration of cerebration as the quintessentially human capacity, mimicry being the highest form of praise. Rather, it reflected a new form of divine worship, an exaltation of the essential endowment of mankind, that unique faculty which man shared with God, because of its link to God, not to man. The thinking machine was not, then, an embodiment of what was specifically human, but of what was specifically divine about humans—the immortal mind.

In Cartesian terms, the development of a thinking machine was aimed at rescuing the immortal mind from its mortal prison. It entailed the deliberate delineation and distillation of the processes of human thought for transfer to a more secure mechanical medium—a machine that would provide a more appropriately immortal mooring for the immortal mind. This new machine-based mind would lend to human thought permanent existence, not just in Heaven, as Kepler imagined, but on earth as well. For its designers, then, the thinking machine unconsciously represented a more perfect "second self," as psychologist Sherry Turkle described it, the reflected glimmer and eternal incarnation of their own divinity.[11]

At first the effort to design a thinking machine was aimed at merely replicating human thought. But almost at once sights were raised, with the hope of mechanically surpassing human thought by creating a "super intelligence," beyond human capabilities. Then the prospect of an immortal mind able to teach itself new tricks gave rise to the vision of a new artificial species which would supersede *Homo sapiens* altogether. Totally freed from the human body, the human person, and the human species, the immortal mind could evolve independently into ever higher forms of artificial life, reunited at last with its origin, the mind of God.

Among the first persons to imagine the possibility of such a thinking machine were the American electrical engineer Claude Shannon and the English mathematician Alan Turing, who together developed the theoretical basis for both the design of electronic computers and the subsequent development of Artificial Intelligence. Confronted by the limitations of mechanical analog computers while overseeing the operations of MIT's Differential Analyzer, the most advanced computation machine of its day, Shannon suggested speeding up and simplifying the system by substituting electromagnetic relays for machined parts, using Boole's binary arithmetic to describe the electrical network. By using the Boolean system, invented to describe the laws of thought, to describe the operation of electric circuits, Shannon laid the groundwork for the electrical simulation of thought—the foundation of electronic computers. He surmised that, if Boole's laws of thought could express the behavior of electronic circuits, electronic circuits could express thought; if the same mathematical terms could be used to describe both human thought processes and the dynamics of an electrical machine, then the two must at least have common characteristics, even if they were not literally identical (a misguided notion provocatively proposed a few years later by the neurophysiologist Warren McCullough and the mathematician Walter Pitts, with their description of the neural networks of the brain in Boolean terms). "Shannon had always been fascinated by this idea that a machine should be able to imitate the brain," Turing's biographer Andrew Hodges noted.[12]

Imitation of the mind was precisely the preoccupation of the eccentric mathematician Turing, who shared Shannon's vision of a thinking machine capable of simulating human thought. Shortly before Shannon published his classic master's thesis on electrical switching circuits, Turing had issued his own theoretical description of an abstract "machine," a universal computer, which operated on the basis of the Boolean system and was capable of expressing logical statements. The operation of the so-called Turing machine was based upon the establishment of a precise relationship between the binary arithmetic of the "machine" and a higher-level symbolic notation, which could be used to simulate thought—an analogy, that is, between the states of the machine and the states of mind.

The appearance of Shannon's work on switching networks confirmed Turing's theoretical speculation, and the two men got together to discuss their common obsession. "They had found their outlook to be the same," Hodges noted. "There was nothing sacred about the brain, and . . . if a machine could do as well as a brain, then it *would* be thinking." Turing made this deceptively modest approach more explicit in another classic paper, published, appropriately enough, in the philosophical journal *Mind*. There he described what he called an imitation test, known thereafter as the "Turing test," in which an interrogator, located in one room, was asked to distinguish between a human being and a machine, both located in another, judging only on the basis of written teletyped responses to his questions. In classic Cartesian fashion, Turing pointed out that "the new problem has the advantage of drawing a fairly sharp line between the physical and the intellectual capacities of a man. . . . The form in which we have set the problem reflects this fact in the condition which prevents the interrogator from seeing or touching the other competitors, or hearing their voices." In the wake of the rapid development of electronic computers, Turing confidently predicted that "in about fifty years' time, it will be possible to programme computers . . . to make them play the imitation game so well that an average interrogator will not have more than a seventy percent chance of making the right identification after five minutes of questioning." The machine's performance would then be deemed "intelligent." "We may hope that machines will eventually compete with men in all purely intellectual fields," Turing concluded.[13]

With his minimalist definition of machine intelligence, Turing had deftly sidestepped philosophical discussions about the actual meaning of mind and thought; his materialist approach dismissed at the outset any discussion of the existence of an autonomous mind or a soul, which had preoccupied Descartes and Boole. (Turing had by this time become an avowed atheist.) By his "imitation principle," "if a machine appeared to be doing as well as a human being then it *was* doing as well as a human being." Yet, in his more extravagant musings and theoretical caveats, Turing perhaps revealed another, albeit submerged, dimension to his thought. In his paper on "Computing Machines and Intelligence," he took the matter of machine intelligence a step further than mere imitation, suggesting that machines might someday be designed whose thinking powers could actually evolve beyond what had been originally programmed. "He was not so much concerned with the building of machines designed to carry out this or that complicated task," Hodges noted. "He was now fascinated with the idea of a machine that could *learn*. It was a development of his suggestion in [his earlier paper] 'Computable Numbers' that the states of a machine could be regarded as analogous to the 'states of mind.' If this were so, if a machine could simulate a brain in the way he had discussed with Claude Shannon, then it would have to enjoy the faculty of brains, that of learning new tricks." "One may hope that this process will be more expeditious than evolution," Turing wrote. "The survival of the fittest is a slow method of measuring advantages. The experimenter, by the exercise of intelligence [in machine design], should be able to speed it up.[14]

Here is some seemingly sober speculation about an entirely new development, transcendence not only of the human body but of human intelligence itself (and hence human control)—a machine modeled upon human intelligence but at the same time autonomous of human intelligence, with the ultimate capability of perhaps surpassing and even supplanting its human counterpart. And what precisely would be the ontological significance of such autonomous machines? In countering what he referred to as the "theological objection" to the design of intelligent machines, Turing mockingly dismissed concern about irreverently usurping divine powers or denigrating the crown of creation. Yet his ironic rejoinders reflect the persistence of deep-seated cultural preoccupations. In designing such machines, as in conceiving children, Turing observed, "we are . . . instruments of His will providing mansions for the souls He creates." "Consolation would be more appropriate" in response to those fearful of jeopardizing mankind's privileged position, he wrote: "perhaps this should be sought in the transmigration of souls"—the transfer, that is, of the souls of men to their machines.[15]

Shortly before he apparently took his own life by eating a cyanide-laced apple, Turing sent four last postcards—"Messages from the Unseen World," he called them—to a friend, which contained cryptic reference to a perhaps abiding faith, despite his fashionable atheism. The first card was lost. On the second he wrote, "The Universe is the Interior of the Light Cone of the Creation," referring to the cosmological theories of Einstein. "Science is a Differential Equation, Religion is a Boundary Condition," he wrote on the third. On the last, the message in verse was more extended and evocative of ancient belief: "Hyperboloids of wondrous Light, /Rolling for age through Space and Time; /Harbour those waves which somehow might, /Play out God's holy pantomime."[16]

Whatever the meaning for Turing of these final reveries, the transcendent significance of his and Shannon's work resonated far and wide in a world attuned to the religion of technology, particularly in the revived apocalyptic atmosphere of postwar America. As was the case with the technologies of space exploration, the quest for Artificial Intelligence proceeded primarily within military arsenals.

Nearly all of the theoretical developments that made possible the design of computers and the advance of Artificial Intelligence stemmed from military-related experience. Shannon's contribution evolved from his work on Vannevar Bush's Differential Analyzer, which was developed for and used primarily by the military. Likewise, Turing's reflections on computing machines derived in some measure from his wartime work decoding German cryptography for the British high command (which he later regarded as his flirtation with sin, much as Oppenheimer later saw his atomic-bomb work). Similarly, John von Neumann's crucial contribution to computer-system design, software programming, emerged from his high-level military work for the Manhattan Project during the war and Cold War strategic planning thereafter. And Norbert Wiener's conception of cybernetics, the use of information theory to design servomechanism-controlled, self-correcting machinery, was primarily the product of wartime developments in automatic gunfire control.

These theoretical contributions, together with the war-spawned advances in electronics, automatic control, and computing machinery which they furthered and reflected, provided the intellectual and material foundation that made possible for the first time the practical development of Artificial Intelligence (AI). The earliest efforts in this direction took place at just the same time as the earliest efforts at manned spaceflight, and under the same military auspices. Pioneer AI researchers were involved in military-sponsored projects on what became known as "man-machine" systems, which aimed to achieve a better match between complex aircraft, anti-aircraft and naval-gun, and radar systems and the human personnel that manned them. Hence, as AI developers Herbert Simon and Allen Newell described it, the researchers "were in a position to observe the analogies between human information processing and the behavior of servomechanisms and computers"—that is, to view the human and mechanical parts of the systems as fundamentally the same. The first human models for Artificial Intelligence, therefore, were pilots, gunners, and radar operators. After the Russian explosion of an atomic bomb in 1949, the early researchers became involved in the development of computer-based strategic air-defense systems. In this context, Allen Newell, the RAND Corporation's training director for the SAGE air-defense system, designed the first use of a computer for symbol manipulation rather than number-crunching, along the lines Shannon and Turing had theoretically described. This historic achievement entailed the simulation of aircraft radar blips, the early warning signals of Armageddon.[17]

At RAND, Newell thereafter teamed up with Simon, a management theorist, to design the first programming simulations of complex human decision-making, the detection of radar signals by radar operators, so that this activity could be taken over by an automatic computer system. In the process they also created programs that simulated the human decision-making required for playing chess and proving theorems in mathematical logic. The military context remained the AI environment throughout its development. MIT's AI pioneers Marvin Minsky and John McCarthy, for example, pursued practically all of their researches for decades, first under the auspices of the Office of Naval Research, and later through their own pipeline to the Department of Defense's Advanced Research Projects Agency (ARPA). The military milieu lent a real-world legitimacy, as well as an urgency, to their research, and reinforced their transcendent tendencies.

Edward Fredkin, another fervent apostle of AI, began his computer work in the navy, in which he participated in the development of the SAGE system. As a civilian, he continued his work under military contract at MIT's Lincoln Labs and then in private industry, ultimately joining MIT in its computer-development program. Like his counterparts in the space program, Fredkin was haunted by the specter of the apocalypse; according to one biographer, "He periodically revises his plans for surviving the nuclear war he believes is imminent." Having become independently wealthy from his industrial activities, Fredkin bought his own Caribbean island and fortified it for survival in the world following a nuclear holocaust. "The world has developed means for destroying itself in a lot of different ways, global ways," said Fredkin.[18]

At the same time, while contributing through his work to the technological arms race of the Cold War, he became convinced that the accelerated advance of Artificial Intelligence was the only salvation for mankind, the means by which rational intelligence might prevail over human limitations and insanity. In this spirit, he taught courses at MIT and Stanford on "saving the world." "The idea was to view the world as a great computer and to write a program [the "global algorithm"] that, if methodically executed, would lead to peace and harmony." He later became preoccupied with what he called "digital physics," grounded upon the notion that the universe itself is a computer, and that our world, operating in accordance with the programming of some celestial intelligence, is God's simulation—"God's holy pantomime."[19]

According to the official creation myth of the Artificial Intelligentsia, AI as a self-conscious technological project was launched in 1956. After they had programmed a digital computer to express symbols in SAGE simulations and chess-playing, Newell and Simon, together with J. C. Shaw, formulated their radically reductionist notion of "information processing systems," and, on that theoretical basis, proceeded laboriously to write programs for their computer which would simulate human thought. (Linguistic theorist Umberto Eco has suggested that AI computer languages are "heirs of the ancient search for the perfect language," the pre-Babel universal language of Adam.)[20]

"The basic point of view inhabiting our work has been that the programmed computer and human problem solver are both species belonging to the genus information processing system," Newell and Simon wrote. "The vagueness that has plagued the theory of higher mental processes and other parts of psychology disappears when the phenomena are described as programs." In this spirit, they developed their Logic Theorist program, designed to prove automatically theorems taken from Russell and Whitehead's *Principia Mathematica*—often described as the first actual demonstration of Artificial Intelligence. The first machine proof of a theorem was accomplished in the summer of 1956, after which Simon excitedly wrote to Bertrand Russell about it. Russell replied sardonically: "I am delighted to know that Principia Mathematica can now be done by machinery. I wish Whitehead and I had known of this possibility before we both wasted ten years doing it by hand. . . . I am delighted by your example of the superiority of your machine to Whitehead and me." (Interestingly, perhaps, at this same time Simon wrote his one and only short story, in which he aimed to illustrate in nonmathematical language the decision-making model of a maze described in his 1956 paper, "Rational Choices and the Structure of the Environment." His story was entitled "The Apple" and it centered upon the protagonist's progressive understanding of the Genesis myth about Eve's temptation in the Garden of Eden.)[21]

That same year, the eventual vanguard of Artificial Intelligence convened for the first time at Dartmouth College, typically considered the founding event of the AI enterprise. The conference was organized by John McCarthy of MIT, who is credited with having named the new field and later established the AI program at Stanford. It was attended by, among others, Marvin Minsky, who went on to direct MIT's AI program; Newell and Simon, who oversaw the AI program at Carnegie Mellon; IBM's Nathaniel Rochester; and Claude Shannon. The explicit aim of the conference was to imagine practical advances toward the advent of "intelligent machines." According to the conference proposal, "the study is to proceed on the basis of the conjecture that every aspect of learning, or any other feature of intelligence, can in principle be so precisely described, that a machine can be made to simulate it." Here Newell and Simon first presented their Logic Theorist and Minsky produced his influential field-defining paper, "Steps Toward Artificial Intelligence." [22]

Perhaps because of his penchant for hyperbole, as well as MIT's inside track to DARPA (Defense Advanced Research Project Administration), Minsky became the premier promoter of Artificial Intelligence. His intentionally provocative denigrations of human mental anatomy and ability gained him widespread notoriety, as did his extravagant exaggerations of AI advances. Beyond his propagandistic motives, his pronouncements showed a deep disdain for mere mortality and an impatience for something more. Minsky described the human brain as nothing more than a "meat machine" and regarded the body, that "bloody mess of organic matter," as a "teleoperator for the brain." Both, he insisted, were eminently replaceable by machinery. What is important about life, Minsky argued, is "mind," which he defined in terms of "structure and subroutines"—that is, programming. Like Descartes, he insisted that the mind could and should be divorced from both the body and the self. "The important thing in refining one's own thought," Minsky maintained, "is to try to depersonalize your interior." The possibility of an utter separation of the mind from the thinking person underlaid his belief in the possibility of a thinking machine—"machines that manufacture thoughts"—and he viewed intelligence as something that could be achieved by any "brain, machine, or other thing that has a mind." [23]

"Can we someday build intelligent machines?" Minsky asked. "I take the answer to be yes in principle, because our brains themselves are machines. . . . Even though we don't yet understand how brains perform many mental skills, we can still work toward making machines that do the same or similar things. 'Artificial Intelligence' is simply the name we give to that research." As evidence of the advancement toward machine intelligence, Minsky described such machine capabilities as searching, pattern recognition, expert systems, automatic theorem-proving, machine vision, and robotics. But he also looked beyond such mundane manifestations of Artificial Intelligence. "Our mind-engineering skills could grow to the point of enabling us to construct accomplished artificial scientists, artists, composers, and personal companions." "Is Artificial Intelligence merely another advance in technology," he mused, "or is it a turning point in human evolution?"[24]

In the short term, Minsky prophesied at the Dartmouth conference, man-machine symbiosis would become the major manifestation of Artificial Intelligence, long before the advent of truly autonomous thinking machines capable of evolutionary advance. Time-sharing computers, he argued, will enable us "to match human beings in real time with really large machines," rendering the machines practical "thinking aids." "In the years to come, we expect that these man-machine systems will share, and perhaps for a time be dominant, in our advance toward the development of Artificial Intelligence."[25]

The development of AI promised to provide "an extension of those human capacities we value most," Pamela McCorduck exulted. "This thinking machine," she explained, "would amplify these qualities as other machines have amplified the other capacities of our body." Accordingly, the U.S. Air Force sought to use high-speed computers to "amplify" or "accelerate" human cognitive processes, in order to bring pilots "up to speed" and thereby ensure optimal use of their high-performance aircraft; the F14 jet fighter, for example, required split-second pilot responses to a rapid, continuous flow of computer-generated information. The human component of the weapons system thus had to be fitted for "real time interactivity" through the computer-based "augmentation of human intellect."

Air Force research into human-machine symbiosis, the so-called pilot-associate project, included studies of voice-actuated computers, computers that respond to the pilot's eye movements, the control of computers by brain waves (known as "controlling by thinking"), and even the direct "hard-wiring" of pilots to computers.[26]

Long before autonomous intelligent machines superseded the human mind altogether, computers could be employed to expand it through the temporary measure of man-machine systems. (This approach was outlined in 1960 by Manfred Clynes in an article on the use of man-machine systems in space exploration, in which he coined the new term "cyborg" to signify the physical integration of cybernetic mechanical systems and living organisms.) At the same time, the military also experimented with new communication systems which simultaneously linked a collectivity of individuals within a single computer system. SIMNET, for example, developed for the simulation of tank maneuvers, created a "virtual community" of eight hundred people—the crews of two hundred tanks. Likewise, the Department of Defense established the ARPANET to link together military-contractor researchers across the country.[27]

The military development of man-machine systems gave rise to both "virtual reality" computer simulation systems (described by Jason Lanier, who coined the term, as "computerized sensory immersion") and "cyberspace" (the term invented by science-fiction writer William Gibson), the on-line world of computer-mediated communication (via the Internet, originally the ARPANET). Enmeshed in computer-based communication and simulation systems, human beings experienced an "enhancement of the senses" and the seemingly infinite extension of their mental powers and reach—delusions of omniscience, omnipresence, and omnipotence that fueled fantasies of their own God-likeness.[28]

In the 1970s, man-machine system researcher Tom Furness left the Air Force to start up the Human Interface Technology Laboratory at the University of Washington, which quickly joined the vanguard of research on virtual reality and cyberspace. There the future began to "take on a luminous dimension," as one researcher described it, in which ritualistic immersion in computer-simulated realities readily evoked the familiar refrains of the religion of technology. "On the other side of our data gloves," the researcher exulted, "we become creatures of colored light in motion, pulsing with golden particles. . . . We will all become angels, and for eternity. . . . Cyberspace will feel like Paradise . . . a space for collective restoration [of the] habit of perfection."[29]

"Our fascination with computers . . . is more deeply spiritual than utilitarian," computer-industry consultant and philosopher Michael Heim has argued, tracing its roots back to the seventeenth-century mathematician and philosopher Gottfried Wilhelm Leibniz. "When on-line, we break free from bodily existence," from our "earthy, earthly existence," and emulate the *viseo dei,* the perspective of God, the " 'all-at-onceness' of divine knowledge." "What better way to emulate God's knowledge," Heim wrote, "than to generate a virtual world constituted by bits of information. Over such a cyber world human beings could enjoy a god-like instant access." Indeed, the designers of one of the earliest civilian computer-communication networks, a community bulletin board created in 1978 for the San Francisco Bay Area, opened their prospectus with the words: "We are as gods and might as well get good at it." As one sociologist described them, these pioneers of cyberspace were charged with a "technospiritual bumptiousness, full of the redemptive power of technology." "Much of the work of cyberspace researchers," she wrote, "assumes that the human body is 'meat'—obsolete, as soon as consciousness itself can be uploaded into the network. The discourse of visionary virtual world builders is rife with images of imaginal bodies, freed from constraints that flesh imposes." "The body in cyberspace is immortal," declared one enthusiast.[30]

The religious rapture of cyberspace was perhaps best conveyed by Michael Benedikt, president of Mental Tech, Inc., a software-design company in Austin, Texas. Editor of an influential anthology on cyberspace, Benedikt argued that cyberspace is the electronic equivalent of the imagined spiritual realms of religion. The "almost irrational enthusiasm" for virtual reality, he observed, fulfills the need "to dwell empowered or enlightened on other, mythic, planes." Religions are fueled by the "resentment we feel for our bodies' cloddishness, limitations, and final treachery, their mortality. Reality is death. If only we could, we would wander the earth and never leave home; we would enjoy triumphs without risks and eat of the Tree and not be punished, consort daily with angels, enter heaven now and not die." Cyberspace, wrote Benedikt, is the dimension where "floats the image of a Heavenly City, the New Jerusalem of the Book of Revelation. Like a bejeweled, weightless palace it comes out of heaven itself . . . a place where we might re-enter God's graces . . . laid out like a beautiful equation."[31]

Despite the intoxicating reveries induced by computer-based virtual realities, man-machine integration is understood by AI advocates to be merely an interim phenomenon along the path toward fully autonomous intelligent systems. As sociologist Sherry Turkle observed, AI enthusiasts believe that eventually "machines will exceed human intelligence in all respects." Intelligent machines might momentarily magnify human capability, but ultimately they will move beyond human capability altogether. Once able to think for themselves, and endowed with "superintelligence," they will break free from such symbiosis with humans and begin to chart their own independent course. And by means of such mind machines, which will eventually eclipse them in evolution, men will transcend their mortality and at last regain their providential powers. "I have a dream to create my own robot," DARPA researcher Don Norman said. "To give it my intelligence. To make it my mind . . . to see myself in it." "So who doesn't?" fellow researcher Roger Schank concurred. "I have always wanted to make a mind. Create something like that. It is the most exciting thing you could do. The most important thing anyone could do."[32]

By transferring their minds to machines, researchers hoped to liberate them once and for all from bodily limitations, so that they might live forever. In his recent history of the field, AI practitioner Daniel Crevier discussed the relationship between AI and religion and argued that AI is consistent with the Christian belief in resurrection and immortality, quoting from Scripture to support the notion of a material (machine-based or bodily) transcendence of the soul.

"Doesn't the materialist view of the mind contradict the existence of an immortal soul?" he asked, and insisted that both the Old and New Testaments imply "that Judeo-Christian tradition is not inconsistent with . . . bodily resurrection in the afterlife." (He cited passages from the prophets Isaiah, Ezekiel, and Daniel as well as St. Paul, and referred also to the accounts of those who had had near-death experiences and described having abandoned their moribund bodies and inhabited another kind of body, invisible yet with a definite structure.) "It is certain that some kind of support would be required for the information and organization that constitutes our minds," Crevier acknowledged, a material, mechanical replacement for the mortal body. But "religious beliefs, and particularly the belief in survival after death, are not incompatible with the idea that the mind emerges from physical phenomena." Christ was resurrected in a new body; why not a machine?[33]

Crevier recounted the discussions of such a possibility that began to surface on the AI grapevine in the 1980s, in particular the idea of "downloading" the mind into a machine, the transfer of the human mind to an "artificial neural net" through the "eventual replacement of brain cells by electronic circuits and identical input-output functions." "This (so far) imaginary process strongly suggests the possibility of transferring a mind from one support to another," and hence the survival of the "soul" after death in a new, more durable, medium. "This gradual transition from carnal existence to embodiment into electronic hardware would guarantee the continuity of an individual's subjective experience" beyond death. Moreover, "the mental powers of the continuing electronic personality could even be enhanced. Thus, AI may lift us into a new kind of existence where our humanity will be not only preserved but also enhanced in ways we can hardly imagine."[34]

The chief prophet of such "postbiological" computer-based immortality was Hans Moravec, a Stanford-trained AI specialist who joined the faculty at Carnegie Mellon and developed advanced robots for the military and NASA. In 1988, his visionary *Mind Children* described in detail how humans would pass their divine mental inheritance on to their mechanical offspring. (His new book is entitled *The Age of Mind: Transcending the Human Condition Through Robots*).[35]

Moravec lamented the fact that the immortal mind was tethered to a mortal body, and that "the uneasy truce between mind and body breaks down completely as life ends [when] too many hard-earned aspects of our mental existence simply die with us." But, he exclaimed, "it is easy to imagine human thought freed from bondage to a mortal body." Just as a computation in process can be transferred from one computer into another, so the same kind of transfer might be achieved from a thinking mind into a computer. "Imagine that a mind might be freed from its brain in some analogous (if much more technically challenging) way," Moravec mused. Mind might thereby be "rescued from the limitations of a mortal body" and passed on to "unfettered mind children."[36]

Moravec described the surgical procedure involved in such a transfer, which entailed linking the neural bundles of the brain to cables connected to the computer. (Crevier found his description "convincing.") "In time, as your original brain faded away with age, the computer would smoothly assume the lost functions. Ultimately your brain would die, and your mind would find itself entirely in the computer. . . . With enough widely dispersed copies, your permanent death would be highly unlikely." (The same replication procedure would also make possible resurrection, since "the ability to transplant minds will make it easy to bring to life anyone who has been carefully recorded on a storage medium.") Thus, in Moravec's view, the advent of intelligent machines will provide humanity with "personal immortality by mind transplant," a sure "defense against the wanton loss of knowledge and function that is the worst aspect of personal death."[37]

Among the members of the AI community, such longings are common. "We're a symbiotic relationship between two essentially different kinds of things," observed Danny Hillis, an AI disciple of Marvin Minsky at MIT, designer of the Connection Machine, a parallel-processing supercomputer, and cofounder and CEO of Thinking Machines, Inc. "We're the metabolic thing, which is the monkey that walks around, and we're the intelligent thing, which is a set of ideas and culture. And those two things have coevolved together, because they helped each other. But they're fundamentally different things. What's valuable about us, what's good about humans, is the idea thing. It's not the animal thing." Like Moravec, Hillis bemoaned the bounds of his mortal existence. "I think it's a totally bum deal that we only get to live 100 years. I think that's awful, that's barely enough chance to sort of get going. . . . I want to live for 10,000 years. . . . I don't see why that shouldn't be possible if I had a better metabolism. . . . If we can improve the basic machinery of our metabolism . . . If I can go into a new body and last for 10,000 years, I would do it in an instant. . . ."[38]

If intelligent machines were viewed as vehicles of human transcendence and immortality, they were also understood as having lives of their own and an ultimate destiny beyond human experience. In the eyes of AI visionaries, mind machines represented the next step in evolution, a new species, *Machina sapiens*, which would rival and ultimately supersede *Homo sapiens* as the most intelligent beings in creation. "I want to make a machine that will be proud of me," Danny Hillis proclaimed, acknowledging the superiority of his creation. "I guess I'm not overly perturbed by the prospect that there might be something better than us that might replace us. . . . We've got a lot of bugs, sorts of bugs left over history back from when we were animals. And I see no reason to believe that we're the end of the chain and I think better than us is possible." The aim for Hillis was not human perfectibility per se but the "optimality of idea evolution"—that is, the advance of the divine element in humanity, by whatever means. "I believe in the soul and the importance of it," Hillis acknowledged. "I believe that there is something fundamentally good about humans. I'm sad about death, I'm sad about the short time that we have on earth and I wish there was some way around it. So, it's an emotional thing that drives me. It's not a detached scientific experiment or something like that."[39]

"I think our mission is to create artificial intelligence," Edward Fredkin boldly declared; "it is the next step in evolution." There have been three great events of equal importance in the history of the universe, he explained. The first was the creation of the universe itself; the second was the appearance of life; and the third was the advent of Artificial Intelligence. The last, according to Fredkin, is "the question which deals with all questions. In the abstract, nothing can be compared to it. One wonders why God didn't do it. Or, it's a very godlike thing to create a superintelligence, much smarter than we are. It's the abstraction of the physical universe, and this is the ultimate in that direction. If there are any questions to be answered, this is how they'll be answered. There can't be anything of more consequence to happen on this planet."[40]

"The enterprise is a god-like one," AI enthusiast Pamela McCorduck observed. "The invention—the finding within—of gods represents our reach for the transcendent." "It's hard for me to believe that everything out there is just an accident," said Fredkin. The universe itself is the product of "something which we would call intelligent." Fredkin thus implicitly viewed the evolution of Artificial Intelligence as a step toward an ultimate resolution between creator and created, a return of mind to its divine origin. Moravec shared the same eschatological vision. "Our speculation ends in a supercivilization," he prophesied, "the synthesis of all solar system life, constantly improving and extending itself, spreading outward from the sun, converting non-life into mind. . . . This process might convert the entire universe into an extended thinking entity . . . the thinking universe . . . an eternity of pure cerebration."[41]

The full-blown transcendent vision of the *idiots savants* of Artificial Intelligence was recently presented by AI guru Earl Cox, an authority on the design of so-called fuzzy logic systems, in his book *Beyond Humanity: CyberRevolution and Future Mind* (co-authored with paleontologist Gregory Paul). Cox argues that exponentially accelerating advances in science and technology have sped up the course of evolution, outdistancing their creators. We are thus already at the twilight of human civilization and the dawn of a new robotic "supercivilization," which will remake the entire universe in its digital image. Happily, he advises, *Homo sapiens* need not be completely left behind, as the dinosaurs were. "Technology will soon enable human beings to change into something else altogether" and thereby "escape the human condition." "Humans may be able to transfer their minds into the new cybersystems and join the cybercivilization," securing for themselves an eternal existence. "This is not the end of humanity," Cox explains, "only its physical existence as a biological life form. Mankind will join our newly invented partners. We will download our minds into vessels created by our machine children and, with them, explore the universe. . . . Freed from our frail biological form, human-cum-artificial intelligences will move out into the universe. . . . Such a combined system of minds, representing the ultimate triumph of science and technology, will transcend the timid concepts of deity and divinity held by today's theologians."[42]

By the 1980s, Artificial Intelligence had given rise to a companion enterprise known as Artificial Life (there was always a certain degree of overlap between the two realms, and some individuals, such as Edward Fredkin and Danny Hillis, identified themselves with both). As computer researchers created machines with ever more computational capacity, they found they could simulate life and evolution as well as intelligence and experience. Artificial Intelligence was a "top down" approach to the creation of machine-based mind, which began directly with the "transfer" of human intelligence to machines. Artificial Life (A-Life) was a "bottom up" approach, which created the artificial conditions in which virtual mathematical "lifeforms" could evolve and from which Artificial Intelligence would eventually "emerge"—as it were, *ex nihilo, in silico*. For A-Life researchers as for their AI brethren, such simulations signaled an advance in evolution, the creation of a superior new silicon species constituted completely of information, the advent of pure mind-life.

As was the case with AI, the theoretical development of A-Life began in the shadow of Armageddon. The generally recognized father of what became A-Life was the mathematician John von Neumann, the "main scientific voice in the country's nuclear weapons establishment." Toward the early end of his life, suffering from terminal cancer, von Neumann earnestly devoted himself to weapons development, advocating the use of nuclear weapons and favoring a preventive nuclear war. At the same time, he began to ponder the fundamental logical similarities between life and machines, and developed the theory of self-reproducing cellular automata upon which A-Life came to be based. Von Neumann himself produced some of the first A-Life programs, but, as hydrogen-bomb mathematician Stanislaw Ulam, von Neumann's friend and an early A-Life theorist himself, eulogized, "he died so prematurely, seeing the promised land but hardly entering it."[43]

Von Neumann's ideas about self-reproducing automata were tested further by Ulam and other visionary mathematicians, including the Cambridge prodigy John Horton Conway, who devised a program called simply "Life." Later Edward F. Moore and Freeman Dyson developed similar ideas about the possibility of creating self-reproducing factories, based upon von Neumann's theory of automata, which could be deployed on other planets. In the same spirit, the most elaborate practical attempt to develop von Neumann's notion was begun at NASA.

In 1980, NASA established a Self-Reproducing Systems Concept Team to explore the possibilities of self-reproducing factories. Their aim was to examine the feasibility of devising machines capable of production, replication, growth, self-repair, and evolution, machines that could be used to colonize the moon and beyond. The team produced several proposals, including one for a "Growing Lunar Manufacturing Facility," and another for "a fully autonomous, general-purpose self-replicating factory to be deployed on the surface of planetary bodies or moons." These seemingly fanciful proposals were actually fashioned in earnest; the team advocated their development and anticipated and fully expected to receive the necessary funding (which went instead to the Star Wars Strategic Defense Initiative). As team leader Richard Laing recalled, "There was the suggestion, if you could just tease money for this self-replicating factory, you would never need money again. You could take over the universe!" The team had some concerns about going ahead with the program, but nevertheless recommended doing so. "We must assume," they cautioned, "that we cannot necessarily pull the plug on our autonomous artificially intelligent species once they have gotten beyond a certain point of development." According to Laing, NASA administrators endorsed and supported the proposals.[44]

Despite the obvious dangers, NASA's nascent A-Life enterprise was continued. One reason for this official encouragement might well have been religious belief, however unconsciously held. The study team's proposals included not only the technical specifications and feasibility estimates of the projects, but also some reflection about their larger significance. The team compared the emergence of this new silicon species to the "emergence and separation of plant and an-imal kingdoms billions of years ago on Earth." Was mankind just a "biological way station" for this superior new species? they won-dered. "Would humankind be seen as nothing more than an evolu-tionary precursor" of such machines? They worried about whether or not the self-replicating machines would have a soul, or think they had a soul. "Could a self-reproducing, evolving machine have a concept of God?" they asked. In the end, they concluded their reflections on an optimistic note, arguing that only through the development of these artificial life-forms, which were "in a very real intellectual and material sense our offspring," would human beings be able to survive into eternity. Toward that end, they envisioned a permanent coexis-tence between the new species and the old, even a merging of the two, through which "mankind could achieve immortality for itself."[45]

The NASA study team insisted that "machine self-replication and growth is a fundamentally feasible goal," and before long the military began to think so too. By the mid-1980s, the U.S. Air Force Office of Scientific Research was underwriting research along the same lines. The center of this effort, which became the "A-Life mecca," was Los Alamos, birthplace of the atomic bomb. There and at the nearby Santa Fe Institute the "promised land" glimpsed by von Neumann came into sharper focus. In 1987, twenty years after the Dartmouth conference launched the enterprise of Artificial Intelli-gence, the first conference on Artificial Life took place at Los Alamos. In the manner of their Dartmouth predecessors, the apostles of A-Life boldly proclaimed their mission. "Artificial Life is the study of arti-ficial systems that exhibit behavior characteristic of natural living systems," they declared. "Microelectronic technology and genetic en-gineering will soon give us the capability to create new life forms *in silico* as well as *in vitro*."[46]

The term "Artificial Life" was coined by Chris Langton, a com-puter hacker who became obsessed with understanding the funda-mental process of life by creating its essence in mathematical form on a computer. He traces his enthusiasm back to a near-mystical experi-ence he had one day while his computer was running John Conway's "Life" program, which prompted some deep reflection on the meaning of "mortality." He was busy with other things and not watching the computer closely when "suddenly he felt a strong presence in the room. Something was there. He looked up and the computer monitor showed an interesting configuration he hadn't previously encountered." "I crossed a threshold then," he remembered. "You had the feeling there was really something very deep here in this little artificial universe and its evolution through time. . . . Could you have a universe in which life could evolve?" "The ultimate goal of artificial life," he later wrote, "would be to create 'life' in some other medium, ideally a *virtual* medium where the essence of life has been abstracted from the details of its implementation in any particular model. We would like to build models that are so life-like that they cease to become *models* of life and become *examples* of life themselves." "Life needs something to live on, intelligence needs something to think on, and it is this seething information matrix which cellular automatas can provide," A-Life pioneer Rudy Rucker explained. "Cellular automatas will lead to intelligent artificial life. If all goes well, many of us will see live robot boppers on the moon."[47]

The 1987 Los Alamos conference institutionalized this obsession with silicon-based intelligent life. Just as Marvin Minsky had outlined the Artificial Intelligence agenda at the Dartmouth meeting, here the agenda of Artificial Life was advanced by enthusiast and promoter J. Doyne Farmer, one of the central figures in the field at Los Alamos. Farmer heralded what science writer Stephen Levy called "the quest for a new Creation." "Within fifty to a hundred years a new class of organisms is likely to emerge. These organisms will be artificial in the sense that they will originally be designed by humans. However, they will reproduce, and will evolve into something other than their initial form. They will be 'alive' under any reasonable definition of the word. These organisms will evolve in a fundamentally different manner than contemporary biological organisms, since their reproduction will be under at least partial *conscious* control. . . . The pace of evolutionary change consequently will be extremely rapid. The advent of artificial life will be the most significant historical event since the emergence of human beings. . . . This will be a landmark event in the history of the earth, and possibly the entire universe."[48]

"*With the advent of artificial life, we may be the first species to create its own successors,*" Farmer emphasized. And with a paternal sensibility reminiscent of Victor Frankenstein, he described mankind's mathematical progeny. "What will these successors be like? If we fail in our task as creators, they may indeed be cold and malevolent. However, if we succeed, they may be glorious, enlightened creatures that far surpass us in their intelligence and wisdom. It is quite possible that, when the conscious beings of the future look back on this era, we will be most noteworthy not in and of ourselves but rather for what we gave rise to. Artificial Life is potentially the most beautiful creation of humanity."[49]

Buttressed by government funding and institutional support, A-Life advocates shared with their Artificial Intelligence colleagues an arrogant impatience with criticism. Farmer, for example, contemptuously dismissed "people irrationally ranting and raving" about the social implications of Artificial Life technology, and insisted that "to shun Artificial Life without deeper consideration reflects a shallow anthropocentrism." "Right now it's kind of nice in a way that A-Life is underground," he told Steven Levy, "because it means we can keep a low profile and just do what we want." (A-Life researcher Norman Packard betrayed the same devil-may-care attitude toward earthly concerns in his interview with Levy. He predicted that the evolution of superintelligent self-reproducing silicon beings is only a couple of generations away and that, yes, their existence would have serious far-reaching consequences for future humanity, but, he concluded, "What the hell, I'm not going to be around.")[50]

If their privileged and protected environment encouraged such cool confidence, their élan derived also from their being self-conscious members of an elite corps of savants on the threshold of discovering the secrets of creation, the latest incarnation of spiritual men enthralled by the religion of technology. "Once we understand the powers of creation in nature, the result will be very, very, very much more powerful than the discovery of the bomb and it will have much wider consequences," A-Life researcher Steen Rasmussen prophesied.[51]

The fundamentally religious ethos of the A-Life research community was described by Stanford anthropologist Stefan Helmreich, who spent time in residence at the Santa Fe Institute and Los Alamos.

He found that, like the monks and saints of earlier centuries, A-Life researchers lived an almost ethereal existence. Engrossed in their work, their material needs met by a service staff, "they can leave their bodies behind," to commune and even identify with their pure-mind computer creations. Many recalled having had, like Langton, "quasi-mystical epiphanies" which enabled them to see "parts of the inanimate world as infused with life," a life with which they could become intimate. "It would be nice to have friends that had a different set of limitations than we do. I would love to have one of my machines be such a friend," said Danny Hillis.[52]

At the same time, the theoretical premises of their work encouraged them to view themselves as basically no different from their mathematical counterparts, and thus once removed from their mortal embodiment. "I view myself as a pattern in a cellular automata world," said one researcher; another defined organisms as computations "of which I remain convinced that I am one. . . . I can't see what else I could be." Moreover, in the imagination of A-Life researchers, if the advent of their new creation signaled the imminent transcendence of mortality, they were themselves involved in this transcendent prospect, not only as initiators but as participants, their sainthood permanently enshrined on a silicon substrate.[53]

Despite all their intellectual iconoclasm and futuristic fantasies, the A-life researchers remained mired in an essentially medieval milieu of Christian mythology. At least some of them are aware of their lineage. "I believe that science's greatest task in the late twentieth century is to build living machines," Rucker said. "In Cambridge, Los Alamos, Silicon Valley, and beyond, this is the computer scientist's Great Work as surely as the building of the Notre Dame cathedral on the Ile de France was the Great Work of the medieval artisan." Even though most of them were professed agnostics or atheists, Helmreich observed, "Judeo-Christian stories of the creation and maintenance of the world haunted my informants' discussions of why computers might be 'worlds' or 'universes,' . . . a tradition that includes stories from the Old and New Testament (stories of creation and salvation)."[54]

Haunted by the lasting legacy of Adam's transgression, Steen Rasmussen tempered his hubris with the persistent feeling "in some way that I am committing a sin by the things I am doing." Helmreich observed that "A-Life scientists invoke normative notions of God and what I refer to as Judeo-Christian cosmology when they speak about artificial worlds," and regularly referred to programmers as gods. "I feel like God; in fact, I am God to the universes I create," said one researcher. Rucker regarded the work of A-Life designers as "divine interventions." Tom Ray designed his program so that alterations in the configuration of life-forms were confirmed by pressing a button labeled "Amen." Peter Todd described how his system evolved "immortal" organisms; others imagined the possibility of achieving "artificial reincarnation." In at least three A-Life systems, including Langton's pioneering loop system, the seed program with which the artificial evolutionary process was commenced was named "Adam."[55]

"The manifest destiny of mankind is to pass the torch of life and intelligence on to the computer," Rucker proclaimed. Yet, three and a half centuries after Descartes first dreamed of releasing the immortal mind from its mortal moorings, A-Lifers were still wrestling with the enigma of the Christian soul, the divinity of man now boldly being passed forward to its mechanical progeny. "I'm not sure yet whether I think that having a soul is a property of all life or only a property of some kind of higher life," Norman Packard told Steven Levy. "I think the cleanest thing would be to say that all living things have a soul and that is in fact the thing that makes them living." Therefore, he surmised, "if you can envision something living in an artificial realm, then it's hard not to be able to envision, at least at some point in the future, arbitrarily advanced life-forms—as advanced as us—therefore they would probably have a soul, too." "That logic is kind of hard to get around," Packard concluded, on a promising note, "and so that tends to make me think that you can have an artificial soul. I wouldn't say an artificial soul, actually, I'd say you can have a soul in an artificial universe. . . . But it would be a real soul." A real soul, just as the first father of computers, Charles Babbage, imagined our "future state": "unclogged by the dull corporeal load of matter which . . . claims the ardent spirit to its unkindred clay."[56]

CHAPTER 11

POWERS OF PERFECTION:
GENETIC ENGINEERING

The pursuit of perfection through the hardware and software of machines was soon extended to the actual "wetware" of life itself, viewed as merely another sort of machine. Having acquired considerable scientific knowledge and technological ingenuity in the creation of mechanical devices designed to enhance and simulate the powers of living beings, the modern magi were now prepared to bring their prowess to bear upon the stuff of life itself, to understand it and, ultimately, to create it anew. "When the divine secret hidden in the germ shall be perfectly unfolded," Edward Bellamy had prophesied, mankind would achieve "the fulfillment of evolution." "When we acquire the ability to interpret the messages of the genome," J. Doyne Farmer wrote a century later, "we will be able to design 'living things.'" Armed with such knowledge, the genetic engineers would strive, first, to restore their true dominion over the creatures of the earth, and hence their divinely ordained role in creation. Second, in turning their newfound powers upon their own kind, they would seek finally to purify the human species of the physical frailties with which it had been cursed, thereby to restore it to its original perfection.[1]

Here too the modern drama dawned with Descartes. Hermetic philosophers and alchemists had long dreamed of uncovering the secret of life and learning how to create life themselves, typically by means of esoteric incantations or incubations of life-giving "menstruum." Thus the legendary Rabbi Low of Prague breathed the name of God into a clay figure to create his celebrated golem, much as God had breathed a soul into the clay of Adam (interestingly, at least three major AI pioneers—von Neumann, Wiener, and Minsky—believed themselves to be direct descendants of Rabbi Low). And Paracelsus had advised incubating semen in blood to create the living homunculus. It was Descartes, however, influenced by the mechanical automata-makers of his day (he was alleged to have made his own robot), who first imagined that the bodies of living things might be merely machines themselves (their souls, of course, were another matter). Descartes "proposed that the bodies of animals be regarded as nothing more than complex machines," John Cohen noted, and thereby became among the first to "extend the idea of a machine into the domain of living organisms." Francis Bacon was similarly inspired. In his utopian *New Atlantis,* Bacon imagined that the mechanically minded fathers of Solomon's House had reclaimed their rightful dominion over the earth and were thereby destined to bring about the restoration of perfection. Possessed of "the knowledge of causes and secret motions of things," they were capable of "enlarging the bounds of human empire, to the effecting of all things possible." Among such possible things, in addition to "the curing of diseases counted incurable," were "the prolongation of life," "the transformation of bodies into other bodies," and the "making of new species." Once understood in mechanical terms, Bacon imagined, life could be renewed and remade.[2]

Just as the seventeenth-century dreams of spaceflight and disembodied minds had to await the twentieth century to be realized, so too did the dream of dominion over life. As it happens, the critical moment in the fulfillment of this last dream coincided almost exactly with that of the others. At the same time in the mid-1950s when the maiden manned-spaceflight program was begun and the agenda of Artificial Intelligence was initiated, the structure and function of DNA was first disclosed—significantly, in the machine-based vocabulary of codes, programs, and information processing. One person who perhaps most clearly comprehended the transcendent significance of this trinity of events was the British life scientist and pioneer X-ray crystallographer J. D. Bernal, whose field contributed mightily to the understanding of molecular structure, including that of DNA. A quarter-century earlier, Bernal had written his first book, a series of essays entitled *The World, the Flesh, and the Devil*. Although Bernal became a Marxist and an atheist, the subtitle of his book suggested rather an older tradition: *An Enquiry into the Future of the Three Enemies of the Rational Soul*.[3]

The three enemies of the rational soul, Bernal argued, are nature, the body, and human desires and emotions. To escape the first, Bernal proposed a flight from the earth and the construction of enormous spherical orbiting space stations ten miles in diameter, new artificial biospheres removed once and for all from the ravages of nature. "Already ambition is stirring in men to conquer space as they conquered air," he exclaimed, "and this ambition . . . as time goes on becomes more and more reinforced by necessity"—unpredictable geological catastrophe and population explosion. (His colleague J. B. S. Haldane had two years earlier similarly proposed space colonization, in a paper provocatively titled "The Last Judgement.")[4]

To overcome the second, he proposed removing the brain from the body and artificially maintaining its function by mechanical means. "Sooner or later man will be forced to decide whether to abandon his body or his life. After all it is the brain that counts," Bernal wrote. Like Bacon, Bernal envisioned the emergence of a future scientific elite; for Bacon they were technological wizards, for Bernal they had become disembodied brains, "transformed men . . . transcending the capacities of untransformed humanity." Freed from the ravages of age and the flesh—for "bodies at this time would be left far behind"—they would become virtually "immortal," experiencing a "continuity of consciousness" in "a practical eternity of existence." As such, they would be perfectly suited for an extraterrestrial existence (as Bernal noted, "The colonization of space and the mechanization of the body are obviously complementary").[5]

"Normal man is an evolutionary dead end," declared Bernal,

anticipating the daring designs and mysticism of future AI enthusiasts. "Mechanical man, apparently a break in organic evolution, is actually more in the true tradition of a further evolution. . . . The new life which conserves none of the substance and all of the spirit of the old would take its place and continue its development. Such a change would be as important as that in which life first appeared on the earth's surface." In the end, Bernal envisioned a time when, "finally, consciousness itself may end or vanish in a humanity that has become completely etherealized, losing the close-knit organism, becoming masses of atoms in space communicating by radiation, and ultimately perhaps resolving itself into light." "That may be an end or a beginning," Bernal wrote, "but from here it is out of sight."[6]

Bernal acknowledged that there might be opposition to this "aristocracy of scientific intelligence," that there would always be "reactionaries . . . warning us to remain in the natural and primitive state of humanity." But he concluded optimistically that "even if a wave of primitive obscurantism then swept the world clear of the heresy of science, science would already be on its way to the stars. . . . The scientists would emerge as a new species and leave humanity behind."[7]

He conceded that the third challenge, that of the "devil"—desire and emotional confusion—was the most difficult to overcome, because it was less susceptible to any technological fix, but he believed that, with the discipline of medieval monks, the disembodied elite would in time subdue also this last enemy of the rational soul that lay in the path of progress. "The cardinal tendency of progress is the replacement of an indifferent chance environment by a deliberately created one," he insisted. "As time goes on, the acceptance, the appreciation, even the understanding of nature, will be less and less needed. In its place will come the need to determine the desirable form of the humanly controlled universe."[8]

With mankind's dominion assured, the next step in this progress, Bernal suggested, was for mankind to discover how "to make life itself," thereby becoming an active participant in creation. Furthermore, he predicted, "Men will not be content to manufacture life. They would want to improve on it." He did not elaborate. Four decades later, in the year of his death, Bernal wrote a preface for a new edition of his first book. Rather than disavow the extravagant imaginings of his youth, he seconded them. "This short book was the first I ever wrote," he explained. "I have great attachment to it because it contains many of the seeds of ideas which I have been elaborating throughout my scientific life. It still seems to me to have validity in its own right." In his preface, Bernal recounted the recent developments that seemed to have fulfilled his prophecy, particularly in the areas of space exploration and machine intelligence, and he returned as well to the theme he had earlier left undeveloped. "Yet, in my opinion," he declared, "the greatest discovery in all modern science has been one in molecular biology—*the double helix*—which explains in physical, quantum terms, the basis of life and gives some idea of its origin. It is the greatest and most comprehensive idea in all science. . . ." Here at last was the key to his imagined next step: the making and improving of life itself. A physicist, Bernal believed that the reliability of scientific knowledge depended ultimately upon its conformity with the physical laws of nature, and his optimism about the prospect of understanding the processes of life reflected the fact that great strides had already been made in reducing the mysteries of life to just such proportions. The clarification of the molecular structure and function of DNA, in particular, represented the culmination of many decades of effort to understand in such physical terms what has been considered, in Arthur Peacocke's phrase, "the most distinctive feature of living organisms, their ability to reproduce."[9]

In reproducing themselves, organisms become engaged in the creation of new life. And new life bears the unmistakable imprint of the old that produced it. Reproduction is thus not simply a random event, but basically a deterministic one, a development by design. An understanding of this precise process of reproduction, therefore, is perhaps at the same time an understanding of the fundamental mechanism of creation itself. Humans have long recognized that offspring bear some resemblance to their parents, but it was only in the nineteenth century that the patterns and processes of such inheritance began to be understood. The father of this new science, which came to be called genetics, was the Moravian Augustinian priest, monk, and abbot Gregor Mendel. A confirmed creationist schooled in the physical sciences, Mendel described in mathematical terms the patterns of inheritance he observed in successive generations of peas. In so doing, he noted that particular visible traits were passed on as discrete, independent, and irreducible units, one from each parent, which might or might not be expressed in any given generation, according to the laws of probability. Mendel's pioneering work thus not only disclosed the statistical laws of heredity, but also suggested that there was some internal physical mechanism of reproduction which produced such patterns.[10]

Mendel published the results of his investigations in 1865, but they went unnoticed for four decades. Just two or three years after their publication, the Swiss chemist F. Miercher isolated an acidic substance from the nuclei of cells of salmon sperm which he called nucleic acid. Eighty years later, the connection was belatedly made between the two discoveries, when nucleic acid was positively identified as the physical material of the gene. In the interim, biologists strove to pinpoint the locus of the internal machinery of heredity, identifying the nucleus as the seat of hereditary material, then a decade later narrowing it down to the chromosomes. Some early researchers speculated that nucleic acid might be the hereditary material, especially after it was discovered that DNA was a major constituent of chromosomes. At the same time, extended experiments with fruit flies confirmed the Mendelian laws of genetics, while the Mendelian unit of heredity was clarified conceptually, but not yet physically, as a formal, irreducible chromosomal entity called the gene.[11]

The influx of physicists into the field of biology in the 1930s signaled the commencement of what Horace Judson dubbed "the eighth day of creation." Their aim, according to their chief patron, Warren Weaver of the Rockefeller Foundation, was "to build a new biology on the bedrock of the physical sciences," an effort Weaver later christened "molecular biology." Weaver, who held firmly to the determinism of classical physics in the face of quantum theory, believed that the molecular foundation of life, once reduced to its physical basis, would likewise prove to be elegant and orderly. Behind this conviction was his Christian belief in a creator God, "the author of the grand design, ultimately responsible for its intricate beauty and for our evolving capacity to recognize the lovely unity that pervades the apparent diversity." "The explanations of science, when traced down, disappear in either fog or assumption," Weaver later wrote in his autobiography. "The explanations of religion, on the other hand, are founded on faith and conviction. Of the two, the second basis seems to me the more satisfying."[12]

The arrival of the Rockefeller-funded physicists, under Weaver's aegis, quickened the quest for a fundamental physicochemical understanding of the hereditary mechanism, from two directions. On one front were those who were chiefly concerned with describing molecular structure, using the techniques of X-ray crystallography developed in England by W. H. Bragg and his son W. L. Bragg. These included Bragg pupils Bernal and W. T. Astbury, who in the late 1930s undertook the structural analysis of proteins and nucleic acids, and the American Linus Pauling, who later discovered the basic structure of the protein molecule.

On another front were physicists originally schooled in atomic quantum theory who were interested primarily in understanding the mechanism of heredity, the physical means by which information was encoded, preserved, and passed on to new life. On a theoretical level, their efforts commenced with the biological speculations of atomic physicist Niels Bohr and one of the founders of quantum mechanics, Erwin Schrödinger. Experimentally, they began with the investigations of Max Delbrück, a student of both Bohr and Schrödinger, who explored the mechanisms of heredity and genetic programming in single-celled bacteria and bacterial viruses (bacteriophages). Delbrück, Salvador Luria, Alfred Hershey, Oswald Avery, and others of this so-called information school ultimately discerned that such single-cell organisms transmitted reproductive information by way of DNA, thus confirming that this was indeed the hereditary material ultimately responsible for the replication of life.[13]

In 1943, in a series of lectures at Trinity College in Dublin, the anti-Nazi émigré Schrödinger proposed a theoretical integration of the insights of these structural and information schools. His lectures were published a year later as the book *What Is Life? The Physical Aspect of the Living Cell,* which became the "ideological manifesto of the new biology." As such, it was the chief inspiration for a new generation of scientists who would achieve such an integration on a practical level as well, among them Francis Crick and James Watson.[14]

Schrödinger sought to understand life in terms of the physical laws of nature and thus as part of a larger whole. "We have inherited from our forefathers the keen longing for unified, all-embracing knowledge," he declared, invoking the fundamentally monotheistic outlook of Western science. "We feel clearly that we are only now beginning to acquire reliable material for welding together the sum-total of all that is known into a whole." In particular, he asked, "How can events *in space and time* which take place within the spatial boundary of a living organism be accounted for by physics and chemistry?" "The obvious inability of present-day physics and chemistry to account for such events," he wrote confidently, "is no reason at all for doubting that they can be accounted for by those sciences. . . . We must not be discouraged by the difficulty of interpreting life by the ordinary laws of physics."[15]

Drawing upon recent research, especially that of Delbrück, Schrödinger proposed that the hereditary "material carrier of life" was an "aperiodic crystal," a solid at once remarkably stable and infinitely versatile. Its very structure, he suggested, contains a "code-script," information for engendering "order from order" which is "the real clue to the understanding of life." "The clue to the understanding of life is that it is based on a pure mechanism," he argued, a kind of "clockwork" that integrates form and function, whose structure is also information. This physical construction of the gene, he reverently observed, "is the finest masterpiece ever achieved along the lines of the Lord's quantum mechanics." The gene is like a "tiny control office" of each cell, "stations of local government dispersed through the body, communicating with each other with great ease, thanks to the code that is common to all of them."[16]

Like Bernal, Schrödinger celebrated such a physical understanding of life, while acknowledging the naïve reservations one might have in "declaring oneself to be a pure mechanism." "The space-time events in the body of a living being which correspond to the activity of its mind, to its self-conscious or any other actions, are," he asserted, "if not strictly deterministic at any rate statistico-deterministic." He realized, however, that such an insight posed a dilemma, a potential contradiction between such determinism and free will. On the one hand, he argued, "my body functions as a pure mechanism according to the Laws of Nature," yet, on the other hand, "I know . . . that I am directing its motions. . . ."[17]

Thus, in his controversial epilogue, Schrödinger returned to "the more traditional and transcendent issue of mind." Like Descartes three centuries earlier (he used Descartes's *"cogito ergo sum"* for his epigraph), Schrödinger resolved this apparent contradiction by locating the "I" in an altogether different dimension from the body, a divine dimension beyond the laws of nature. "The only possible inference from these two facts is, I think, that I—I in the widest meaning of the word, that is to say, every conscious mind that has ever said or felt 'I'—am the person, if any, who controls 'the motion of the atoms' according to the Laws of Nature," which "in Christian terminology" is equivalent to "Hence I am God Almighty." In earthly experience, I am merely the canvas upon which a variable assortment of data momentarily collects, he concluded. Such experience comes and goes, former selves fade in time, yet "in no case is there a loss of personal existence to deplore. Nor will there ever be." The person, then, is at once physical and spiritual, mortal and immortal, natural and supernatural, a living creature bounded by earthly laws and yet also a divine agency above and beyond them. Schrödinger acknowledged that some might find such religious suggestions embarrassing or even blasphemous, "but please disregard these connotations for the moment," he urged his reader, "and consider whether the above inference is not the closest a biologist can get to proving God and immortality at one stroke."[18]

A decade later, inspired by Schrödinger's manifesto, James Watson and Francis Crick joined forces in Cambridge, bringing the structural and information schools together in practice to decipher the structure and thus the code of DNA. The American Watson was a molecular biologist who had studied with Salvador Luria of Delbrück's "phage group." The English physicist Crick (together with Rosalind Franklin and Maurice Wilkins of London's King's College and some of his own colleagues from the Cavendish Laboratory at Cambridge) brought the technology of X-ray crystallography to bear upon "the heart of a profound insight into the nature of life itself." In eighteen months they had discerned the double-helix structure of DNA and fathomed the physical mechanism of inheritance. "We have found the secret of life," Crick exclaimed. ("And now the announcement of Watson and Crick about DNA," said Salvador Dalí, in words later used by Crick as the epigraph for his *Of Molecules and Man.* "This is for me the real proof of the existence of God.") In an understated way, Watson concurred: "DNA molecules, once synthesized, are very, very stable," he noted. "The idea of the genes' being immortal smelled right." "The double helix has replaced the cross in the biological analphabet," said pioneer nucleic-acid chemist Erwin Chargaff. In time, as Dorothy Nelkin observed, this view of DNA as an eternal and hence sacred life-defining substance—indeed, the new material basis for the immortality and resurrection of the soul—became a modern article of faith. DNA spelled God, and the scientists' knowledge of DNA was a mark of their divinity.[19]

But understanding the structure of DNA in reality represented only the beginning of a physical understanding of the processes of reproduction and inheritance, the "secret of life." For DNA, however important, actually creates nothing; it is merely the physical bearer and conveyor of genetic information according to which new life is to be built. The actual building is done by the myriad mechanisms of the cell as a whole, which read the encoded instructions and produce the specified building blocks of life: amino acids and, from these, enzymes and other proteins, including, of course, DNA itself. Thus, once the message-bearing structure of DNA was known (and the precise "sequencing" of DNA bases was undertaken), the tedious task of understanding these complex cellular activities was begun in earnest. If the advocates of Artificial Intelligence viewed machines as potentially lifelike, the molecular biologists viewed life as essentially machinelike. Just thirty years after the publication of Schrödinger's speculative manifesto, they could declare with confidence that the supposed mystery of life was indeed, at bottom, just another mechanism. "We are looking at a rather special part of the physical universe which contains special mechanisms none of which conflict at all with the laws of physics," molecular biologist Sidney Brenner observed matter-of-factly in 1974. "That there would be new laws of Nature to be found in biological systems was a misjudged view and that hope or fear has just vanished."[20]

In his youthful futuristic vision of 1929, J. D. Bernal had observed that scientists would not be content with understanding or even making life; "they would want to improve on it." The actual manufacture of life from scratch remained a distant and largely unspoken dream (according to Lewis Mumford, one departing president of the American Chemical Society, a Nobel laureate, exhorted his colleagues in 1965, "Let us marshal all our scientific forces together in order to create life!"). But the controlled growth and "improvement" of life soon became a distinct possibility. "My guess is that cells [including those of humans] will be prepared with synthetic messages within twenty-five years," Marshall Nirenberg, who won a Nobel Prize for his role in deciphering the "language" of the genetic code, predicted in 1967. Once the new science had given rise to an adequate technology, biology, the study of life, would increasingly become at the same time biotechnology, the engineering of life.[21]

By 1970, having outlined the genetically controlled processes of the cell in single-celled organisms like bacteria, the molecular biologists began to move on to the larger and infinitely more complex cells of higher organisms. They learned how to isolate such cells, taken from embryos, for investigation. More important, they learned how to remove DNA fragments from these higher-organism cells and insert them into the plasmid and chromosomal DNA of bacteria in order to study their effects in a less complex and more familiar environment and, in the process, identify the sites and functions of particular genes. Crucial to these developments was the discovery of various enzymes which naturally manipulate nucleic acids, copying DNA strands, breaking them up and reassembling them. Armed with these enzymes, the molecular biologists were now able to manipulate the genetic material themselves, precisely removing particular pieces and splicing them together with others to create an entirely new set of genetic instructions. By 1974, these techniques of "recombinant

DNA," first developed in the American laboratories of Herbert Boyer, Stanley Cohen, Paul Berg, and other molecular biologists, had become the basis for genetic engineering.[22]

In addition to its purely scientific value for the study of the genetic mechanisms, genetic engineering proved to have considerable commercial, and therapeutic, significance. Once the higher-organism genes for the production of particular proteins were inserted into bacteria, these bacteria would become in essence a living factory for the rapid reproduction of these genes and the prolific manufacture of these proteins. Moreover, as it happened, the products, processes, and even the genes themselves could be patented, which rendered such technical development extremely lucrative. The pharmaceutical industry quickly recognized the potential of this "biofacture." Natural bovine growth hormone, for example, artificially produced in great quantities in bacteria containing bovine genes, was thereafter administered to cows to increase their milk output. Before long, bacteria (and yeast) were likewise being fitted with human genes to produce, among other things, human insulin and human growth hormone, for therapeutic use in humans.[23]

In his *New Atlantis*, Francis Bacon envisioned that the scientific fathers of Solomon's House would be capable of producing "new species," chimeras which combined the traits of different kinds of creatures. In so doing, Bacon believed, they would decisively demonstrate not only the restoration of mankind's rightful dominion over all other creatures but also its God-like participation in the process of creation itself. Armed with their powerful genetic technology, molecular biologists were soon putting their fictional forebears to shame, conducting experiments that went far beyond natural cross-breeding practices. By learning how to transfer genes from the fertilized cells of one species into those of another, they devised a genetic shortcut through sexual reproduction, directly creating new "transgenic" beings which eventually blurred the boundaries between species and between the plant and animal kingdoms.[24]

By inserting genes from flounder into plants, they rendered the plants frost-resistant. In a similar manner they engineered plants that were hardier or more easily harvested or that produced their own internal nitrogen fertilizer. They inserted human growth genes into the fertilized eggs of pigs and cows to yield farmyard giants. They grew mice endowed with human cancer genes to create cancer-prone strains designed exclusively for laboratory experimentation. They placed insect-resistant genes from tobacco plants into sheep, growth genes from cattle and chicken into salmon and trout, and even genes for fluorescence from fireflies into tobacco plants. In the same spirit, they experimented with the use of milk-producing animals, as they had earlier with bacteria, for the manufacture of human proteins. Researchers at the University of California gave the world a true chimera, the "geep," a cross between a goat and a sheep, which had the body of a sheep but the face and horns of a goat. And, finally, researchers in the United States and Scotland have succeeded in producing mammalian clones, exact genetic duplicates, from embryo cells in monkeys and adult cells in sheep.[25]

But if the new technology endowed bioengineers with Adamic dominion and God-like powers over nature, enabling them to "improve" upon presumably lesser living organisms according to their own lights, needs, and interests, it also, and perhaps most important, enlarged the prospect for their own, human, perfection. The same technology was soon being employed to improve upon the genetic inheritance of human beings as well as plants and other animals.

By the mid-1990s, medical researchers in human genetics had identified and isolated the "defective" genes that were responsible for a range of inherited diseases, including sickle-cell anemia, Tay-Sachs disease, cystic fibrosis, neurofibromatosis, Huntington's disease, and ADA (adenosine deaminase) deficiency. Such information (by 1995, some three thousand alleged genetic diseases had been identified) was quickly put to use, as genetic diagnoses, the screening of people for defective genes, genetic counseling, and prenatal "preventive" screening all became common practices. But identifying defective genes, informing people of their presence, counseling them about appropriate procreative planning, screening them from some high-risk activities and environments, and aborting affected fetuses, merely set the stage for actual biotechnological intervention, in the form of "gene therapy."[26]

Gene therapy went beyond merely identifying a defective gene to actually correcting that defect through genetic engineering. In 1980, Martin Cline, a genetic engineer who had experimented with animals (he had successfully inserted foreign genes into mice), decided to move on to humans. In a highly controversial and ultimately unsuccessful experiment, he injected genetically engineered cells into several female patients to treat blood disorders. His unauthorized experiment triggered a furor over the ethical implications of such practices, the extension of techniques developed in plant and animal genetic engineering, to humans. Nevertheless, gene therapy proceeded apace.

The "first legally sanctioned gene engineering experiment on humans" took place in May 1989, when medical geneticist W. French Anderson and his colleagues from the National Institutes of Health's genetic-engineering team conducted their "gene-marker" experiment. The controversial experiment entailed injecting radioactive genetic "markers" into immune cells taken from terminally ill cancer patients and transfusing the cells back into the patients' bodies in order to track them and monitor their function. This procedure was not meant as therapy but only as research, since no cure was expected or intended.[27]

In September 1990, Anderson and his team performed the first officially sanctioned somatic (body) cell gene-therapy experiment on a four-year-old girl afflicted with ADA deficiency (a severe immune disorder popularly known as "bubble-boy syndrome"). They transfused the girl with cloned animal retrovirus into which the missing ADA gene had been inserted; the retrovirus entered the girl's somatic cells, deposited the needed gene, and thereby corrected the deficiency. Her condition improved, and Anderson declared the experiment a dramatic success. It later turned out, however, that the girl had also received other drug therapy which had proved successful in such cases, thus calling into question the true effectiveness of the gene therapy and raising questions about its propriety, especially in light of the possible risks (the retrovirus used to carry the gene into the girl's body cells was later found to cause cancer in primates).[28]

Some medical genetic engineers expressed alarm at Anderson's audacity. Arthur Bank, professor of medicine and human genetics at Columbia University, for example, described Anderson's procedures as "absolutely crazy" and charged that the NIH researchers were driven by ambition, not science. (Three years earlier, Anderson had cofounded a human gene-engineering company, Genetics Therapy, Inc.) "The main impetus [for the ADA experiment]," Bank argued, "is the need for French Anderson to be the first to do gene therapy in man." Harvard Medical School's Stuart Orkin observed that "a large number of scientists believe the experiment is not well founded scientifically. . . . I'm quite surprised that there hasn't been more of an outcry against the experiment by scientists who are completely objective." Medical geneticist Richard Mulligan, the only member of the NIH Recombinant DNA Advisory Committee to vote against the experiment, said, "If I had a daughter, no way I'd let her get near these guys if she had that defect." Nevertheless, despite the controversy, Anderson emerged as the foremost practitioner of gene therapy, and by 1994 over a dozen such experiments were in process throughout the world, with many more proposed protocols already in the pipeline.[29]

Somatic gene therapy, whether performed upon a person or upon an unborn fetus (or even an embryo), was intended to correct for catastrophic genetic deficiencies, to offer a normal life to those genetically predestined to an early doom. The effects of such therapy were restricted to the individual. But in plant and animal experimentation, genetic manipulation included germline engineering (the altering of the genetic material of germ cells or, more commonly, justfertilized embryos), thereby projecting the effects of genetic engineering onto future generations. By 1995, some observers were expecting that germline experiments would soon be performed upon humans, such as people who were diagnosed as passive carriers of a dangerous defective gene.

Though human gene therapy was presumably intended to restore infirm individuals to health by correcting for a defective inherited endowment, it also held out the promise of "enhancement" of the already healthy. Small-statured people have already taken genetically engineered human growth hormone to correct their "deficiency." Some day, geneticists such as Penn State's Robert Plomin imagine, slow thinkers might be able to have their cognitive faculties upgraded by an infusion of intelligence genes. In his prophecy of 1929, J. D. Bernal acknowledged the need to recruit the "aristocracy of scientific intelligence" from the ranks of ordinary people, with uncertain results. This would be necessary, he wrote, "until we can know from the inspection of an infant or an ovum that it will develop into a genius." In 1990, the NIH awarded a major grant to geneticist Plomin to track down the genes for IQ, in order to identify those whom Plomin described as "the really smart kids."[30]

But in the vision of the vanguard of genetic engineering, the genetic enhancement of present individuals would represent only a prelude to the eugenic engineering, and perfection, of their progeny. "Can we develop so sound and extensive a genetics that we can hope to breed, in the future, superior men?" the Rockefeller Foundation's Warren Weaver had asked in 1934. The geneticist Hermann J. Muller early encouraged eugenic selective breeding, proposing sperm banks for human specimens and genetic control through artificial insemination. In 1939, Muller, together with twenty-two other distinguished geneticists, issued a "Geneticist's Manifesto" for their eugenic agenda. "A more widespread understanding of biological principles will bring with it the realization that much more than the prevention of genetic deterioration is to be sought for," they declared, "and that the raising of the level of the average of the population nearly to that of the highest now existing in isolated individuals, in regard to physical well-being, intelligence and temperamental qualities, is an achievement that would . . . be physically possible within a comparatively small number of generations." Thus everyone might look upon "genius . . . as his birthright."[31]

Thirty years later, in 1969, at the dawn of the genetic-engineering revolution, the distinguished molecular geneticist Robert Sinsheimer proclaimed "a new eugenics" which went far beyond selective breeding. "The old eugenics was limited to a numerical enhancement of the best of our existing gene pool," Sinsheimer explained. "The new eugenics would permit in principle the conversion of all the unfit to the highest genetic level." "It is a new horizon in the history of man," he declared. "Some may smile and may feel that this is but a new version of the old dream, of the perfection of man. It is that, but it is something more. . . . To foster his better traits and to curb his worse by cultural means alone has always been, while clearly not impossible, in many instances most difficult. . . . We now glimpse another route—the chance to ease the internal strains and heal the internal flaws directly, to carry on and consciously perfect far beyond our present vision this remarkable product of two billion years of evolution."[32]

In the same paper of 1950 in which he presented his celebrated test for machine intelligence, mathematician and AI visionary Alan Turing entertained a vision of a different sort: the cloning of human beings. In describing his "Turing test," he tried to define carefully what he meant by a machine, to distinguish it clearly from a living, thinking mind. He noted that some might suggest simply that the machine must be designed by engineers of only one sex, thereby eliminating the possibility that it would be the living product of procreation. But Turing suggested (alert to the prospect perhaps because he was himself homosexual) that it would one day be possible for individuals of either sex to reproduce without the aid of the other, simply by cloning themselves anew from any one of their own body cells—"a feat of biological technique deserving of the highest praise." (Because of the ambiguity produced by this possibility, Turing insisted that to qualify for his test the machine must be an electronic digital computer.) In 1993, barely a half-century later, such human cloning had become a reality. Jerry Hall and Robert Stillman of the George Washington University In Vitro Fertilization and Andrology Laboratory announced that they had indeed successfully cloned human embryos in their laboratory. The eugenic implications were widely understood. Bernard Davis of the Harvard Medical School indicated that he would prefer to see especially the cloning of those individuals, like Turing himself, who excel "in fields such as mathematics or music where major achievements are restricted to a few especially gifted people."[33]

Any kind of human genetic engineering, whether for curative, enhancement, or eugenic purposes, depended ultimately upon an identification of the chromosomal site, internal sequence (the precise base order of their DNA), and function of particular genes. Only after genes for particular traits and diseases had been located could they be isolated, cloned, and used in therapy. Up until the 1990s, the identification of genes had been a rather haphazard affair, the result of independent inquiries by widely dispersed investigators with varying priorities. In the mid-1980s, however, some leading American researchers began to lobby for the establishment of a coordinated, comprehensive, federally funded effort to "map" and "sequence" the entire "human genome," consisting of all one hundred thousand genes that make up a human being. The effort began with Robert Sinsheimer, then president of the University of California at Santa Cruz, who presided over an initial conference on the human genome in 1985. "For the first time in all time," Sinsheimer declared, "a living creature understands its origin and can undertake to design its future." He later made explicit the religious significance of this transcendent scientific endeavor.[34]

"Throughout history, some have sought to live in contact with the eternal," Sinsheimer explained. "In an earlier era, they sought such through religion and lived as monks and nuns in continual contemplation of a stagnant divinity. Today, they seek such a contact through science, through the search for understanding of the laws and structure of the universe and the long quest back through time and evolution for our own origins." "Perhaps this urge is a riposte to fate, a nay to human mortality," he added. "I am a scientist, a member of a most fortunate species. The lives of most people are filled with ephemera. . . . But a happy few of us have the privilege to live with and explore the eternal."[35]

In this spirit, Sinsheimer explained the transcendent significance of the Human Genome Project. "From the time of the invention of writing, men have sought for the hidden tablet or papyrus on which would be inscribed the reason for our existence in this world. . . . How poetic that we now find the key inscribed in the nucleus of every cell of our body. Here in our genome is written in DNA letters the history, the evolution of our species. . . . When Galileo discovered that he could describe the motions of objects with simple mathematical formulas, he felt that he had discovered the language in which

God created the universe. Today we might say that we have discovered the language in which God created life. . . . After three billion years, in our time we have come to this understanding, and all the future will be different."[36]

The year after the Santa Cruz conference, Department of Energy physicist Charles DeLisi sponsored another conference on the human genome at Los Alamos, where he had already set up the GenBank computer system to collect data on DNA sequencing. Like Sinsheimer, DeLisi compared the significance of the new venture with that of the Manhattan Project and the space program, and it was here, at the birthplace of the atomic age, that the molecular geneticist and Nobel laureate Walter Gilbert first proclaimed the human genome the "grail of human genetics."[37]

After an intense lobbying effort by Walter Gilbert, James Watson, Charles Cantor, Leroy Hood, and other leading figures in genetic engineering, together with lobbyists from pharmaceutical and biotechnology companies, the Human Genome Project was established by the U.S. government. It went into formal operation in 1990, under the direction of Watson. Estimated eventually to cost three billion dollars, the project was the largest engineering undertaking since NASA's Apollo Project. In 1993, in a dispute over the patenting of human genes (which Watson opposed), Watson resigned and was succeeded by University of Michigan medical geneticist Francis Collins, a Gilbert protégé, who had contributed significantly to the identification of the genes for cystic fibrosis, neurofibromatosis, and Huntington's disease.

The establishment of the Human Genome Project, with its high-level political support, ample funding, central coordination, research centers throughout the country (including at Lawrence Livermore and Los Alamos), a veritable army of coordinated researchers, and an extensive network for international collaboration, signaled as never before that the era of human genetic engineering had begun in earnest. There was even a sense of urgency on the part of the elder architects of the project. "There is a greater degree of urgency among older scientists than among younger ones to do the human genome now," wrote Watson. "The younger scientists can work on their grants until they are bored and still get the genome before they die. But to me it is crucial that we get the human genome now rather than twenty years from now, because I might be dead then and I don't want to miss out on learning how life works." "This is truly the golden age of biology," Leroy Hood declared. "I believe that we will learn more about human development and pathology in the next twenty-five years than we have in the past two thousand." Walter Gilbert said that in this mighty undertaking he beheld a "vision of the grail." And in the same spirit that had sent legendary medieval knights in search of the most coveted and mysterious prize of Christendom, project director Francis Collins pronounced the unprecedented effort "the most important and the most significant project that humankind has ever mounted."[38]

On the whole, the development of human genetic engineering was no doubt fueled, consciously or not, by enduring medieval myths of artificially engendering human life. Tales of the golem and the elusive alchemical elixir of life, of magically bestowing life upon dead matter, were told and retold, while allusions to their modern scientific equivalent, Mary Shelley's *Frankenstein*, abounded. And the image of the homunculus, the motherless child of hermetic lore, hovered over the accelerated parallel efforts at artificial reproduction—through *in vitro* fertilization and embryo transfer (the animal-tested technical bases for "test-tube babies" and surrogacy) and experimental advances toward the artificial womb (using lambs and goat fetuses)— which contributed enormously to genetic research. (The techniques of *in vitro* fertilization and embryo transfer especially were essential for experiments in germline genetic manipulation. Indeed, as Janice Raymond has argued, helping the infertile was more a rationale—and marketing strategy—than a reason for such developments. Molecular biologist Erwin Chargaff acknowledged that the alleged demand for such reproductive methods on the part of infertile couples "was less overwhelming than the desire on the part of the scientists to test their newly developed techniques," and Raymond noted that "Chargaff's view is supported by reports that over 200,000 embryos have been stockpiled in European IVF centers that have been specifically created for research.")[39]

All such imaginings, as Gilbert's allusion to the "grail" suggested, reflected deeper religious roots. (With regard to Gilbert's invocation of the "grail"—which became emblematic of the Human Genome Project—population geneticist Richard Lewontin has suggested that "it is a sure sign of their alienation from revealed religion that a scientific community with a high concentration of Eastern European Jews and atheists have chosen for its central metaphor the most mystery-laden object of medieval Christianity." More to the point here, it is a sure sign of the enduring influence of the mythology of medieval Christianity in the shaping of Western consciousness, to which these individuals too are heir, whether or not they are Christians themselves.) According to the dominant Judeo-Christian male creation myth of divine descent, the male God created Adam and gave him life (a feat Rabbi Low, with God's help, repeated) unaided by either woman or sex. And God created Eve from Adam, not Adam from Eve (promoting—and reflecting—fantasies of masculine birth and the homunculus). And God created Christ *through* Mary but not *of* Mary (making her the first surrogate mother). Such myths of exclusively paternal, and divine, procreation inspired the earnest endeavors of (preponderantly male) bioengineers, promising them not only a womb of their own, but divine powers of creation as well.[40]

Thus, in their own professional personas and anointed activities, no less than in the holy alphabet of DNA itself, genetic engineers labored in the presence of God. They imagined that, with their new insights into the mechanisms of life, they had come closer than ever before to share in divine knowledge, and with their new technical capabilities for manipulating the basic material of life, they had in a sense become God's companions in creation.

Among the members of one distinguished molecular-biology laboratory, for example, allusions to the godliness of their work were common during laboratory discussions. According to a sociologist of science who spent a year as a participant observer in that lab, the researchers there were often guided in their interpretations by considerations of what God might have been up to; "God wouldn't have done that," they would say about an inelegant interpretation. "They believed they had an inside track, privileged access to divine knowledge, which they identified with knowledge of DNA," the sociologist observed. Graduate students in the lab were required to build their own models of the double helix by way of initiation into the sacred mysteries of their trade. DNA models became the icons of the laboratory, symbols of a divine presence, shrines to be worshipped in silence and wonder. In the shadow of such monuments to perfection, and caught up in the monastic dedication inspired by them, the pioneers of genetic engineering passionately pursued their calling.[41]

And if they labored in the presence of God, the biotechnologists labored as well in the name of God. Inspired in part by the evolutionary eschatological vision of proponents of "process theology"— preached, for instance, by the influential geneticist and evolutionary biologist Theodosius Dobzhansky—the genetic engineers came to believe that they enjoyed divine sanction for their work, that they had been doubly blessed, first, in having been granted God's "gift" of genetic knowledge, and, second, in having been given a role, as the image-likeness of God, in the evolutionary process of creation itself.[42]

The belief in mankind's image-likeness to God, and hence his license to perform godly acts, took two forms, the bold form of "co-creation," or the somewhat more humble form of "stewardship." The first was given its most forceful expression by the Cambridge University biochemist Arthur Peacocke, who had had, as a researcher on the solution properties of DNA, a "grandstand seat" at the birth of molecular genetics. Peacocke noted in one of his theological books that the idea of man's being a co-creator as a consequence of his divine image-likeness (and heavenly destiny) goes back to the Christian humanist movement of the Italian Renaissance. According to historian C. E. Trinkhaus, this movement gave rise to "an important new conception of man as actor, creator, shaper of nature and history, all of which qualities he possesses for the very reason that he is made 'in the image-likeness' [of God]." Thus, some Renaissance thinkers believed that "man's ingenuity and inventiveness is so great that man himself should be regarded as a second creator of the natural world." Marsilio Ficino, for example, "could not help seeing in man's mastery of the world further evidence of man's similarity to God if not of his divinity itself." "Man acts as the vicar of God," wrote Ficino, "since he inhabits all the elements and cultivates all, and present on earth, he is not absent from the ether."[43]

Following in this tradition, Peacocke argued that "man now has, at his present stage of intellectual, cultural, and social evolution, the opportunity of consciously becoming *co-creator* and *co-worker* with God in his work on Earth, and perhaps even a little beyond Earth." In assuming this exalted identity, Peacocke insisted, mankind could avoid the nemesis of hubris—the sin of Adam—"by virtue of his recognition of his role . . . as auxiliary and co-operative rather than as dominating and exploitative." "Man, with his new powers of technology and with new knowledge of the ecosystem," Peacocke said, "could become that part of God's creation consciously and intelligently co-operating in the processes of creative change. . . . The exploration which is science and its progeny, technology, might then . . . come to be seen as an aspect of the fulfillment of man's personal and social development in co-operation with God who all the time is creating the new. Man would then, through his science and technology, be exploring with God the creative possibilities within the universe God has brought into being. This is to see man as *co-explorer* with God."[44]

Most explicitly religious proponents of genetic engineering have eschewed the co-creator image in favor of that of stewardship, without in any way discounting divine sanction for their scientific and technological enterprise. This belief is widely held across the spectrum of the Christian Church, but perhaps the chief proponents of this view among genetic engineers are those associated with the American Scientific Affiliation (ASA), an evangelical Christian organization of some three thousand scientists based in Ipswich, Massachusetts. All members of the ASA must sign a doctrinal statement of faith in which they agree to "accept the divine inspiration, trustworthiness, and authority of the Bible in matters of faith and conduct" and identify themselves as "stewards of God's creation."[45]

One of the most prominent members of the ASA is the director of the Human Genome Project, Francis Collins. A born-again Christian, Collins has always been and remains outspoken about his religious beliefs. "If in fact, God is real, if Christ really walked the earth, if he really hung on the cross as a means of providing a bridge by which we can have direct access to God," Collins has argued, "then that is the most important event in all of history, and upon that rests our present existence and our future in the afterlife. If one comes to the conclusion that that is true, then to shy away from talking about it is, perhaps, to commit intellectual suicide." In the manner of NASA Apollo Project leader Wernher von Braun, Collins insists that "there is no conflict between being an absolutely rigorous scientist and being a person of faith," although he acknowledges that his belief in the "supernatural" poses some difficulties for him as a scientist. "The basic way I look at it is that as soon as you accept the possibility of the supernatural—which of course you can never prove or disprove by the natural—then there is no reason that it has to at all times follow natural laws. . . . I do think that the historical record of Christ's life on Earth and his Resurrection is a very powerful one. And I don't have problems with God intervening from time to time. . . . It doesn't feel like a suspension of my role as a scientist to believe in the ability of the Almighty to break those rules when he sees fit. . . ." As a steward of God, fulfilling God's continuing plan for creation, Collins appears quite comfortable at the head of one of history's most ambitious technological ventures. "There is only one human-genome program," he told a reporter upon his appointment as director. "It will only happen once, and this is that moment in history. The chance to stand at the helm of that project, and to put my own personal stamp on it, is more than I could imagine." "The work of a scientist involved in this project, particularly a scientist who has the joy of being a Christian," Collins explained to a religious conference in 1996, "is a work of discovery which can also be a form of worship." It provides, moreover, a privileged entrée to divine knowledge. "As a scientist, one of the most exhilarating experiences is to learn something, to understand something, that no human understood before—but God did."[46]

Donald Munro, the director of the ASA, is himself a geneticist and physiologist as well as an evangelical Christian. In his view, the recent developments in genetic science and technology constitute a "gift from God" which extends mankind's "dominion" over nature and better enables it to fulfill its "stewardship" function. He is concerned about the possible abuse of this gift by the scientific community, in the face of financial and professional pressures, and concedes that most people find the notion of genetic enhancement "scary," but he remains confident that gene therapy will prove a blessing for mankind if used in an "enlightened" way. He is excited about the prospects of gene therapy for overcoming such "defects" as myopia and mental retardation, as well as fatal diseases, and views preventive genetic screening as a way of reducing the number of abortions. The potential for perfection is infinite, he believes, echoing Robert Sinsheimer's insight that "perfection is like a rainbow that recedes as we approach it." Cautious but committed, Munro is optimistic about the future. "God holds the future," Munro maintains. "God will ensure that we don't go too far afield."[47]

ASA members have written widely on the moral implications of genetic engineering, stressing the notion of mankind's stewardship to God. Like Munro, Hessel Bouma, a widely cited medical geneticist from Calvin College, expressed serious caveats about the possible abuse of the new technology, especially in the direction of eugenics, but nevertheless strongly supported further gene-therapy development. In the same spirit of stewardship, V. Elving Anderson, emeritus professor of genetics at the University of Minnesota, entitled his book (co-written with Bruce Reichenbach in 1994) *On Behalf of God*.[48]

As stewards of God, "made in God's image and likeness," Elving Anderson explained, "we are not simulating a divine role. . . . We are carrying out the divine mandate." Following the prescriptions given Adam in the first two chapters of Genesis, he defined the responsibilities of God's stewards as "filling, subduing, and caring for" creation. "A stewardship ethic sees technology as a gift," he wrote, by means of which mankind can fulfill its divine mandate, but, he cautioned, "we are to act on behalf of God, not out of human hubris." Anderson raised many of the caveats about the potential abuse of genetic technology but went further than most observers in endorsing the genetic alteration of human germline cells as well as somatic cells, and supporting genetic enhancement along with gene therapy.[49]

"Scientists would emerge as a new species and leave humanity behind," Bernal had written in 1929. In the same spirit, Anderson imagined a utopian future—though not one without its own ethical dilemmas—in which a genetically selected and enhanced scientific elite, carrying a cargo of equally screened and altered frozen embryos (together with artificial wombs), abandon the ecologically crippled earth and embark, via their orbiting space station, upon an effort to colonize Mars. "The earth does not need more humans," Anderson wrote, "but perhaps it needs better humans, humans more disease-resistant, genetically superior, more intelligent, sympathetic, moral, and spiritual, better adjusted to and able to cope with their environment. With our rapidly increasing knowledge about the human microsphere and our developing technology, we stand in a position to improve our progeny."[50]

"Already we can diagnose and treat diseases before birth and perform fetal surgery," he pointed out. "Through an analysis of a couple's genetic load, we can predict the probability that their children will inherit certain genetically determined characteristics and use this information in genetic counseling. We possess the knowledge and ability to determine the genetic structure of embryos in vitro, so that the physician can implant in the uterus only those free of genetic defects that would result in painful diseases or life-threatening deformities. If we develop the capacity to perform germline genetic intervention, we might be able genetically to tailor future generations to certain broad specifications." "A qualitative interpretation of the injunction [to fill the earth]," Anderson concluded, "appears to give us the permission—and perhaps more strongly, since it is a command, the obligation—to change the creation for the better. In the past we have focused on changing the environment for human betterment. Now we have enormous powers to begin to redesign the kinds of human beings we want on earth."[51]

The sanction of divine stewardship, albeit interpreted in a more restrained manner, is invoked by more moderate Christian supporters of genetic engineering as well as ardent evangelicals. This was the dominant position espoused, for example, by what might be called the "official" religious arm of the Human Genome Project. Centered at the Institute of Religion of the Texas Medical Center in Houston, the effort is directed by J. Robert Nelson, author of an influential book entitled *Genetics and Religion*. Funded by the Human Genome Project itself, via the good offices of C. Thomas Caskey—director of the Institute of Molecular Genetics of Baylor College of Medicine and president of the International Human Genome Organization—the institute held a series of high-profile conferences in 1990 and 1992 on the topic "Genetics, Religion, and Ethics." Assisting Nelson in directing the program was Hessel Bouma, and prominent among the keynote speakers were Francis Collins and French Anderson. Not surprisingly, Nelson's book, which "grew out of" the two conferences, was, despite its even-handed and cautionary tone, a thinly veiled religious apology for the Human Genome Project.[52]

The distinguished geneticist Caskey was designated the "principal investigator" for the federally funded conferences and lent it official endorsement. "Those who are skeptical about the value of religious truth may feel the need to listen to the new dialogue between genetic scientists and theologians," Caskey wrote in the foreword to Nelson's book. He might have added "those who are skeptical about genetic engineering," the real focus of concern for the religious apologists, who aimed, it seemed, to marginalize or co-opt such skeptics, and divert attention from the issues they raised. Often inspired by their own religious beliefs, critics of genetic engineering typically used similar language of stewardship—interpreted as the divine mandate to preserve and protect nature and humanity—to condemn certain practices, such as human germline manipulation and the patenting of life. (The vast majority of churches endorsed the Human Genome Project itself, however.) In 1995, a coalition of religious groups called for a ban on the patenting of human genetic material by those involved in the Human Genome Project (an issue that had earlier prompted Watson's departure from the directorship). Director Francis Collins decried the critics' campaign, describing the patenting issue as a complex legal rather than a simple moral matter and warning that such a misguided effort would diminish the credibility and reputation of the Christian Church.[53]

Collins's relatively relaxed attitude on the issue of gene patenting perhaps reflected his religious views, which actually minimized the ultimate significance of genetic material. Many critics shared with geneticists the view that DNA and genes represented the essence of life and were thus somehow sacrosanct (although this did not stop most geneticists from groping for a piece of the commercial action). Collins, on the other hand, though he exhorted his students at the University of Michigan to "love DNA," looked elsewhere for the human essence. For him, as for Schrödinger, the material manifestations of life were of little significance compared with the spiritual— what he described as "the part of us that's connected with the eternal and the supernatural." The same was the case for his entrepreneurial colleague, French Anderson.[54]

In November 1991, Anderson delivered a talk at a conference at the National Cathedral in Washington (which he later repeated at the second Houston conference) entitled "Can We Alter Our Humanness by Genetic Engineering?" Anderson recounted the triumph of his ADA-deficiency experiment and attempted to evaluate the ultimate consequences of manipulating "the very core of our being." Was there the danger that we would somehow thereby distort or diminish our humanness, our defining essence? He rhetorically philosophized about what that essence might be, attempting to define and quantify the traits that characterize us as human. Disclosing his own belief in "a supernatural Being," and a "resurrected soul," he concluded that there was no cause for concern, because humanness resides not in the body at all but in the "soul"—"that unmeasurable dimension that is not dependent on the physical hardware of our bodies, . . . that non-quantifiable, spiritual part of us that makes us uniquely human." "If what is uniquely important about humanness is not defined by the physical hardware of our body," Anderson argued, "then since we can only alter the physical hardware, we cannot alter that which is uniquely human by genetic engineering. . . . We cannot alter our soul by genetic engineering." However much we might manipulate the physical, material components of our living beings, therefore, our essence survives untouched—"the uniquely human, the soul, the image of God in man."[55]

Anderson's conviction about the supernatural essence of humanity provided him with a rationale for downplaying the dangers of gene therapy. (At the same time, it bolstered his opposition to genetic enhancement, since no measure of manipulation of the body could in any way improve upon the perfection already present in the soul.)

Most genetic engineers, however, continued to act as if their physical enterprise was indeed a project of perfection, as if their accumulated knowledge and techniques might ultimately restore mankind to its pristine condition, freed from the myriad debilitating defects inherited from the Fall. "The perfection theme is very strong among Human Genome Project participants," observed sociologist Sheldon Krimsky, a former NIH Recombinant DNA Advisory Committee member. "They tend to view the human genome as being riddled with imperfection, with defects, and their aim is to perfect it." (In 1993, a new company was formed in Cambridge, Massachusetts, to commercialize the latest advances in genetics research. Founded by internationally respected figures in the field, the company quickly became a world leader in the brave new world of gene-based therapeutics. As if to highlight their perfectionist ambitions, the founders named their new company "Millennium.")[56]

But since every person has a unique genome, different by roughly three million DNA base pairs from any other, the question is often raised: What exactly is *the* human genome? Whose genome is it? The answer given to this question by Human Genome Project researchers betrays the persistence of a familiar religious conception of human perfection. "They all said that the first human genome to be mapped and sequenced would not be the genome of any particular person but a composite," historian Daniel Kevles observed, recounting his extended conversations with the main architects of the Human Genome Project, including James Watson, Walter Gilbert, Leroy Hood, and Charles Cantor. "The first complete human sequence was expected to be that of a composite person," Kevles wrote in the introduction to a project anthology he edited with Leroy Hood. "It would have both an X and a Y sex chromosome, which would formally make it a male. . . . He would be a multinational and multiracial mélange, a kind of Adam II. . . ."[57] And if the mapping of the human genome charted the terrain of human perfectibility, the project of perfection would finally be fulfilled through human cloning, the asexual reproduction of human life, by design, enabling man to create man, as God had created Adam, in his own image.

THE POLITICS OF
PERFECTION

The millenarian promise of restoring mankind to its original God-like perfection—the underlying premise of the religion of technology—was never meant to be universal. It was in essence an elitist expectation, reserved only for the elect—the "happy few," in Robert Sinsheimer's felicitous (Shakespearean) phrase. Half the species, women, were expressly excluded (see appendix), and so too were the vast majority of the male population, who would likewise be left behind by the saints. Thus the cloistered monks—the spiritual soldiers of salvation epitomized by Joachim of Fiore's millenarian vanguard of *viri spirituales*—pursued their own privileged perfection far in advance of the rest of humanity, as did the mendicant friars who followed in their footsteps, as missionaries and schoolmen. The great explorers too believed that they alone had been chosen, and sent, to rediscover paradise, and the hermetic philosophers and learned magi they so inspired were similarly assured of their own special monopoly on divine wisdom. Stirred by the apocalyptic visions of just such an elite brotherhood of pious wise men, the scientific virtuosi of the seventeenth century imagined themselves the blessed new saviors of mankind, best prepared by their studies and knowledge to meet again in the glorious kingdom to come. And the mantle of perfection they so proudly wore, woven by monks, was passed on, through the closed ranks and secret rituals of Masonic society, to the enlightened elite of modern civilization, the engineers.[1]

But the elite, other-worldly pretensions of all those who promoted and pursued the perfectionist religion of technology were belied by their worldly dependence and subordination. For it was ultimately from worldly power, which they served to enlarge and extend, that their own privileged position, and luxury to dream, derived. Thus Erigena first philosophized about the religion of technology while serving as court philosopher to the Carolingian monarch Charles the Bald, who fought for control over the crumbling empire created by his grandfather Charlemagne. And it was under Carolingian auspices, and in its service, that the Benedictine orders first gained their true terrestrial might and spiritual authority. Thereafter their privileges depended upon their fealty to feudal lords and lay kings, and, ultimately, upon their obedience to the papacy.

Though they were among the first to elevate the useful arts by lending to labor the dignity of worship, the Benedictines soon relegated the real work of their prosperous abbeys to their lay brothers, servant sisters, and peasant wage workers, while they devoted themselves exclusively to the liturgy, the scriptorium, and the garden. By the tenth century, for the Benedictine monks of Cluny, as Jacques Le Goff has pointed out, "labor was exalted mainly in order to increase the productivity and docility of the laborers."[2]

The same transformation subsequently overtook the Cistercians as well, the righteous reform Benedictines who had condemned the Cluniac corruption of monastic ideals. "The poor monks who once maintained themselves through manual labor," George Ovitt observed, "became feudal lords who supervised the work of others." Provided for by "an increasing manual labor force of serfs and wage earners," they amassed great wealth and enjoyed the privileges and prerogatives of the elite, in the service of popes and princes. Thus the Cistercian abbot Joachim of Fiore, though the author of a revised millenarianism which later fueled medieval rebellion, "was not consciously unorthodox and had no desire to subvert the Church. It was with the encouragement of no less than three popes that he wrote down the revelations with which he had been favored." (Indeed, as Bernard McGinn wrote, the millenarianism which Fiore inspired "was as often a rallying cry for the defense of the established order as it was a form of revolutionary ideology. . . . The evidence suggests that the most important and effective innovations in apocalyptic ideas were usually not the products of semi-educated renegades . . . but were produced by the establishment intelligentsia of the day." This was even more the case in the seventeenth-century millenarian revival.)[3]

Heirs to monasticism, the mendicant friars likewise owed their institutionalized existence and prestige to the papacy, which they served, in pious attendance to repression and conquest, with unprecedented diligence and dedication. The friars were monks who had abandoned the cloister in order to evangelize the world. As academic scholars, they laid the intellectual foundations for papal authority as well as science, and as missionaries they lent religious sanction and support to papal, and later imperial, expansion. In the process, they encountered the myriad menaces to established power—the imagined army of Antichrist—which they warned about and warred against. Thus the Franciscan Roger Bacon, an early enthusiast of technological advance, proposed his prescient project of invention to popes, urging that "the church should consider the employment of these inventions against unbelievers and rebels. . . ."[4]

Columbus, of course, pursued perfection in the name of God, guided by the words of prophets. But he did so with the support of, and in fealty to, the Spanish monarchs, for whom he plundered the promised land. The Renaissance magi labored at learning and mastered their magic in pursuit of divine knowledge, only to share their secrets for a sum with the royal patrons who underwrote their efforts. The Rosicrucians, first trumpets of scientific sainthood, bound their terrestrial fortunes to the ill-fated monarchy of Bohemia.[5]

The savants of the seventeenth-century scientific revolution steered a similar course. Francis Bacon dreamed of a New Atlantis but devoted his life's energies to the enrichment of the royal court. In his vision of the millennium, as Margaret Jacob emphasized, Bacon "always located control of leadership in the millennial paradise firmly in elite hands." Likewise, in worldly affairs he sought to enlarge human dominion over nature while preserving intact the established order. In an age of incessant social instability, as James R. Jacob observed, "science [became] another means, along with work discipline and the reformation of manners, by which European elites, having distanced themselves from the people, [sought] to control and subject them to authority." Bacon himself disdained what he called the "innate depravity and malignant disposition of the common people." He exhorted his peers to learn from lowly artisans, not to emulate them but only to enhance their more exalted efforts. Galileo displayed a similar disdain for "women and ordinary folk"—"the shallow minds of common people"—and he urged the Church to hide from the people scientific truth about the heavens lest they become confused and troublesome. Bacon believed, however, that science would teach "the peoples [to] take upon them the yoke of laws and submit to authority, and forget their ungovernable appetites. . . ."[6]

Bacon's followers sustained this elitist outlook. Though they earnestly envisioned the advent of an earthly millennium, Hartlib, Dury, Plattes, and other early Baconians depended heavily upon parliamentary power and privileges and held to a rigidly hierarchical view of society. In their educational-reform efforts, for example, they promoted universal education but divided their schools into "mechanical" and "noble": the first to educate the masses in practical matters, the latter to educate the elite in theory and advanced science. The scientific societies which emerged in the seventeenth century, modeled upon Bacon's vision, owed their existence and allegiance to royal authority, and aristocratic (and increasingly capitalistic) patronage. Accordingly, they viewed their social mission in much the same way as Bacon had his. The Royal Society thus pooled "talents and interests in order to benefit the elite and not the people," argued James Jacob, "in order indeed to contain and exploit the people by drawing upon their knowledge and skills, while at the same time deflecting them from political and religious courses that threatened constituted authority."[7]

The mechanical philosophy, and especially the Newtonian system, served both church and state by providing a seeming naturalistic buttress to the inviolability of the established order. This legacy of the scientific revolution was perpetuated in the eighteenth century by the Newtonian Freemasons, science-minded aristocrats who combined mysticism and magic with a "dedication to order, hierarchy, and perfectibility." In the nineteenth century, it attained its fullest expression in the positivist philosophy of Auguste Comte—and in the person of the engineer who embodied it. Designed deliberately to counter the French Revolutionary tradition, Comte's engineering approach to society was aimed above all at the permanent re-establishment of order. "The motto that I have put forward as descriptive of the new political philosophy," Comte wrote, "is 'Order and Progress.' " "In all cases," he added, "considerations of progress are subordinate to those of order." Whether that order was to be achieved on behalf of a state or a capitalistic enterprise, or both, the engineers, in just the manner Comte envisioned, remained devoted to that end.[8]

In their elite obeisance and service to established power, the twentieth-century proponents of the religion of technology have outdone their predecessors. The engineers of nuclear weapons, endowed from the outset with the authority and limitless largesse of the state, have devoted their energy and imagination to an enlargement of state power. And their counterparts and colleagues in space exploration have done likewise. Von Braun aimed for the stars but hit London and Antwerp, on behalf of the Third Reich. Later he prepared for future terror on behalf of the American armed forces. Throughout most of their career, the men who built the U.S. (and Soviet) space programs served military ends; in their quest for space travel they brought the world but minutes away from mutually assured annihilation. Thereafter, under the nominally civilian auspices of NASA, they have continued to contribute to the militarization of space, in terms of both surveillance capability and the capacity for weapons deployment.

In the same setting, the pioneers of Artificial Intelligence, in quest of the immortal mind, have been sustained by the U.S. military—together with their disciples in Artificial Life, cyberspace, and virtual reality. As they have trained their minds for transcendence, they have contributed enormously to the world arsenal for warfare, surveillance, and control. And they also have placed their technological means at the disposal of manufacturing, financial, and service corporations, which have deployed them the world over to discipline, deskill, and displace untold millions of people, while concentrating global power and wealth into fewer and fewer hands.

Finally, the genetic engineers, supported by the state, have laid the technological foundations for an Orwellian future. At the same time, they have turned their technical prowess to profitable advantage, becoming consultants, shareholders, and directors of entrepreneurial biotechnology and multinational pharmaceutical firms involved in the wholesale patenting and monopolization of plant, animal, and even human "life-forms." Moreover, the profit-spurred acceleration of genetic experiments has made health, safety, ecological integrity, and biological diversity mere secondary considerations, and the routine, unregulated production and utilization of human genetic information has added yet a new means to the arsenal of social discrimination. The long-range eugenic implications of this knowledge and technology, viewed in the light of twentieth-century experience, are neither obscure nor unimaginable.[9]

In all of these areas, the other-worldly preoccupations of latter-day spiritual men have produced unprecedentedly powerful means toward worldly ends. The technologists' expectation of restored dominion has been indulged by their patrons in the interest of enlarged domination. Yet, for the most part, lost in their essentially religious reveries, the technologists themselves have been blind to, or at least have displayed a blithe disregard for, the harmful ends toward which their work has been directed.

When people wonder why the new technologies so rarely seem adequately to meet their human and social needs, they assume it is because of the greed and lust for power that motivate those who design and deploy them. Certainly, this has much to do with it. But it is not the whole of the story. On a deeper cultural level, these technologies have not met basic human needs because, at bottom, they have never really been about meeting them. They have been aimed rather at the loftier goal of transcending such mortal concerns altogether. In such an ideological context, inspired more by prophets than by profits, the needs neither of mortals nor of the earth they inhabit are of any enduring consequence. And it is here that the religion of technology can rightly be considered a menace. (Lynn White, for example, long ago identified the ideological roots of the ecological crisis in "the Christian dogma of man's transcendence of, and rightful mastery over, nature"; more recently, the ecologist Philip Regal has likewise traced current justifications of unregulated bioengineering to their source in late-medieval natural theology.)[10]

As we have seen, those given to such imaginings are in the vanguard of technological development, amply endowed and in every way encouraged to realize their escapist fantasies. Often displaying a pathological dissatisfaction with, and deprecation of, the human condition, they are taking flight from the world, pointing us away from the earth, the flesh, the familiar—"offering salvation by technical fix," in Mary Midgley's apt description—all the while making the world over to conform to their vision of perfection.[11]

But it is not the practitioners alone who are so moved. A thousand years in the making, the religion of technology has become the common enchantment, not only of the designers of technology but also of those caught up in, and undone by, their godly designs. The expectation of ultimate salvation through technology, whatever the immediate human and social costs, has become the unspoken orthodoxy, reinforced by a market-induced enthusiasm for novelty and sanctioned by a millenarian yearning for new beginnings. This popular faith, subliminally indulged and intensified by corporate, government, and media pitchmen, inspires an awed deference to the practitioners and their promises of deliverance while diverting attention from more urgent concerns. Thus, unrestrained technological development is allowed to proceed apace, without serious scrutiny or oversight—without reason. Pleas for some rationality, for reflection about pace and purpose, for sober assessment of costs and benefits—for evidence even of economic value, much less larger social gains—are dismissed as irrational. From within the faith, any and all criticism appears irrelevent, and irreverent.

But can we any longer afford to abide this system of blind belief? Ironically, the technological enterprise upon which we now ever more depend for the preservation and enlargement of our lives betrays a disdainful disregard for, indeed an impatience with, life itself. If dreams of technological escape from the burdens of mortality once translated into some relief of the human estate, the pursuit of technological transcendence has now perhaps outdistanced such earthly ends. If the religion of technology once fostered visions of social renovation, it also fueled fantasies of escaping society altogether. Today these bolder imaginings have gained sway, according to which, as one philosopher of technology recently observed, "everything which exists at present . . . is deemed disposable." The religion of technology, in the end, "rests on extravagant hopes which are only meaningful in the context of transcendent belief in a religious God, hopes for a total salvation which technology cannot fulfill. . . . By striving for the impossible, [we] run the risk of destroying the good life that is possible." Put simply, the technological pursuit of salvation has become a threat to our survival.[12]

The thousand-year convergence of technology and transcendence has thus outlived whatever historical usefulness it might once have had. Indeed, as our technological enterprise assumes ever more awesome proportions, it becomes all the more essential to decouple it from its religious foundation. "Transcendence is a wrong-headed concept," Cynthia Cockburn has argued. "It means escape from the earth-bound and the repetitive, climbing above the everyday. It means putting men on the moon before feeding and housing the world's poor. . . . The revolutionary step would be to bring men down to earth." But respite from our transcendent "faith in the religion of the machine," as Lewis Mumford long ago insisted, requires that we "alter the ideological basis of the whole system." Such an undertaking demands defiance of the divine pretensions of the few in the interest of securing the mortal necessities of the many, and presupposes that we disabuse ourselves of our inherited other-worldly propensities in order to embrace anew our one and only earthly existence.[13]

APPENDIX

A MASCULINE MILLENNIUM:
A NOTE ON TECHNOLOGY AND GENDER

Persistent efforts by women in recent years to breach the so-called traditionally masculine bastions of science and technology have consistently proved less than successful. In order to understand why, it might be helpful to learn how these vital fields of human endeavor became traditionally masculine in the first place, and how the history that shaped them continues to haunt and hamper such efforts. In an earlier book, *A World Without Women*, I tried to account for the gendered construction of science by tracing the ideology and institutions of Western science to their roots in the celibate, misogynist, and homosocial clerical culture of the Latin Church, and to suggest that the legacy of this lineage persists in today's scientific milieu.

I want now to suggest that the religion of technology described here might help us to account for the powerful cultural affinity between technology and masculinity in Western society. For, if the religion of technology elevated the arts, it at the same time masculinized them. By investing the arts with spiritual significance and a distinctly transcendent meaning, the religion of technology provided a compelling and enduring mythological foundation for the cultural representation of technology as a uniquely masculine endeavor, evocative of masculinity and exclusively male. Insofar as the technological project was now aimed at the recovery of Adam's prelapsarian perfection, the image-likeness of man to God, it looked back to a primal masculine universe and forward to the renewal of that paradise in a masculine millennium.

Adam signified the ideal of restored perfection, and that ideal was male. So too were the apostles of the religion of technology, the successive generations of monks, friars, explorers, magi, virtuosi, Masons, and engineers. And so too are their ideological descendants who have designed the hallmark technologies of our own age and given the name "Adam" to the first manned spaceflight, the seed programs of Artificial Life, and the composite human genome. Of course, women might participate, but only marginally at best, because, by definition, they could never aspire to, much less hope to achieve, the ultimate transcendent goal.

In reality, women have always been actively involved in the actual advancement of the useful arts, contributing daily and significantly to the practical activities of human sustenance, security, and survival. As technology came to be mythologically defined as masculine, however, their presence, efforts, and achievements became ideologically invisible. "Women invent, but are not . . . recognized as inventors. This . . . is the whole of the story," observed Autumn Stanley, author of the first encyclopedic study of women's historic and enduring contributions to the development of the useful arts. Stanley has amply documented the full range of female invention from the dawn of human society to the present age, and has concluded, "Women invent. Women have always invented. . . . Women still invent. They invent significant things. They create breakthroughs and fundamental inventions. . . . And they do all this in the full range of human endeavor and technology." "The real question," she argues, "is not, why so few? but why do we know so few?" The exclusive identification of technology with men, on the one hand, and the invisibility of women as agents of technological development, on the other, are, she insists, but reverse sides of the same cultural coin: "the stereotypes separating women and technology." The religion of technology contributed significantly to the creation of such stereotypes.[1]

As late as the Middle Ages, the useful arts were identified as much with women as with men, and women were engaged in almost all aspects of technological practice. Indeed, it was in part because of that female association that the arts were disdained and disregarded by elite men. Carolingian legislation refers to "women's workshops" for the making of linen, wood products, wool combs, soap, oils, and vessels. In a twelfth-century description of the crafts, women were identified not only as weavers and spinners but also as metalworkers and goldsmiths. Parisian guild regulations of the thirteenth century refer to female apprentices, and even craft masters, particularly in the silk and woolen trades. Moreover, the distaff, primarily a woman's tool, was emblematic not only of women's work but of the useful arts and productive labor in general. If some technologies were traditionally identified with men—especially those relating to hunting, warfare, toolmaking, and metalworking, and the ornamental arts associated with religion and state power—others were identified with women; as Ivan Illich has argued, technological activities, including the use of specific tools, were traditionally divided between gendered domains.[2]

In short, the totality of the useful arts belonged to neither domain. Likewise, though guild regulations often specified male hegemony, they at the same time acknowledged respected roles for daughters, wives, and widows of guild members. Men dominated the craft, but they never altogether defined it. Finally, as women steadily lost ground to men, for numerous reasons (including politicized guild regulations, new social legislation, extended markets, the increasing separation of public and private spheres, the diminished importance of household production, the exclusion of women from educational institutions, etc.), their role was relatively diminished in many crafts, although women remained, and even increased their numbers, in others. But this meant merely that men became predominant in certain areas, not that the useful arts per se became totally male. The actual participation and status of women relative to men in the arts, therefore, do not by themselves account for the emergence of so exclusive an ideological identification of technology with men. The relative exclusion of women from the arts did not cause, but more than likely followed from, the cultural representation of technology as uniquely masculine, an extreme and totalistic notion that reflected rather the rise of the religion of technology. In short, the alleged exclusive "masculinity" of technology historically had no reference in reality (at least up until the relatively recent monopolization of the useful arts by professional engineering), only in mythology. It was a mythic rather than a social construct, but one with profound social implications.[3]

The ideological masculinization of the useful arts and the ideological elevation of the useful arts were two sides of the same coin, and both were the product of the belated association of the most humble and worldly of human activities with the other-worldly spirit of transcendence. For it was only when the arts came to be invested with spiritual significance that they became worthy of the attention of and identification with elite males, and the specific Adamic content of that spiritualization reinforced that identification.

Throughout recorded history, men had monopolized the transcendent realms, through their exclusive identification with the ritualized activities of hunting, warfare, religion, and magic. The arts related to such activities—especially metalworking and goldsmithing—had also been associated with the transcendent. Now, for the first time, this transcendent realm was extended to encompass the useful arts in general. At the heart of this change was a renewed emphasis in the early-medieval West, a time of significant advance in technology, on its core monotheistic Judeo-Christian male creation myth, whereby men consciously sought to imitate their male god, master craftsman of the universe, either directly, by assuming a new God-like posture vis-à-vis nature, or indirectly, through a reassertion of their image-likeness to God. The latter reflected a renewed identification not only with Christ—the mythic male Son of God, the last Adam, who symbolized the promise of redemption and the prospect of new beginnings—but also with the first Adam, the mythic first man, whose original but lapsed image-likeness to God inspired efforts toward its recovery.[4]

Just as the Judeo-Christian story of the creation and Fall betrays a decidedly masculine bias, so the recovery of mankind's image-likeness to God was understood by orthodox Christians from the outset to be restricted to males—those whom Augustine called "the sons of promise." God is the male Father of the universe, who creates a son in his image, and it is this masculine divine image which is lost and restored.[5]

In the first chapter of Genesis there is some ambiguity on this point ("So God created man in his own image, in the image of God created he him; male and female created he them"—Genesis 1:27). But however much heterodox commentators used this passage to assert a positive female role in the story of redemption, orthodox commentary, which became the dominant interpretation in the West, either ignored it or treated it allegorically (Augustine interpreted "female" and "male" in a spiritual rather than corporeal sense and argued that the former signified the Church, the latter Christ). In what became the dominant interpretation of the creation story, Church fathers referred instead either to the preceding passage ("Let us make man in our image, after our likeness"—Genesis 1:26) or to the quite different account of creation in the second chapter of Genesis—which thus became far more familiar—in which Adam is created before Eve. Here Adam receives the breath of life directly from God, whereas Eve is created from Adam. (In Christian iconography, God's role in the creation of Eve becomes more remote over time. Initially God is seen removing Adam's rib and transforming it into Eve. In medieval representations, however, God has become a mere midwife, removing a fully formed Eve from the side of Adam, who has, in effect, given birth to her—a procreative reversal common to male creation myths.) Here only Adam, the male, was created in the image of God. This was made explicit by Paul in his first letter to the Corinthians, in which he insisted that women who pray or prophesy must cover their heads, whereas men who do likewise should not: "For a man indeed ought not to cover his head, forasmuch as he is the image and glory of God: but the woman is the glory of the man" (I Corinthians 11:7).[6]

Thus Eve did not share in the original divine likeness. Indeed, it was because of her that Adam, and all the sons of Adam thereafter, lost their divine likeness. According to the fathers of the Church, woman, through her vulnerability to Satan and her temptation of Adam, brought about the Fall and destroyed man's original perfection. "You are the devil's gateway," Tertullian wrote of woman.

"You desecrated the fatal tree, you first betrayed the law of God, you softened up with your cajoling words the man against whom the devil could not prevail by force. The image of God, Adam, you broke him as if he were a plaything." Thus did woman bring desolation and death to mankind and take from man his once-exalted role in creation. Because of her he lost his immortality, his share in divine knowledge, and his divinely ordained dominion over nature.[7]

If Eve did not share in original perfection, neither could she lose or recover it: the restoration of perfection was a project for men only. According to the Book of Revelation, the guidebook for two thousand years of such expectation, the possibility of resurrection in the millennium is restricted to those "which were not defiled with women, for they are virgins" (Revelation 14:4). As one recent commentator on this passage noted, not only does it indicate the importance of chastity, or at least continence, but it is "expressed . . . from an exclusively male point of view."[8]

As woman was the proximate cause of the Fall and hence of the loss of man's original perfection, so she remained the perpetual impediment to its recovery. Thus, insofar as the useful arts came to be viewed as a vehicle of such a recovery, they were deemed to be, by definition, for men only, just as the presence of women was, by definition, perceived as antithetical to the entire project. For the restoration of perfection was a male-only pursuit, an exclusively masculine means back to a primordially masculine beginning: Eden before Eve.

The pursuit of the masculine millennium began within a culturally contrived world without women, a celibate monastic environment which prefigured the promised return of this primordial patriarchal paradise. (Ernst Benz described celibacy as "an anticipation of impending perfection.") This masculine milieu had its origins in the rise of monasticism from the fourth through the sixth centuries, but it had lost much of its ascetic rigor and gender purity in the centuries thereafter. In the ninth century, however, under the imperial auspices of the Carolingians, the spirit of monasticism underwent a thoroughgoing reform and revitalization and became institutionalized as a social force as never before, its ethos extending beyond the cloister into the imperial court itself. It was thus during the so-called Carolingian renaissance that men, through the power of the reformed imperial state, were able to monopolize many social spaces formerly shared with women, from the monasteries themselves to the rarefied realms of higher learning. The Carolingian sponsors of such efforts at strict sexual segregation were also avid supporters of development in the useful arts, and it was under their protection, in the writings of court philosopher John Scotus Erigena, that the ideological transformation of the useful arts began.[9]

Erigena inhabited a world without women, a single-sex environment which was reflected in his contemplation on the spiritual significance of the useful arts. In his revision of Capella's allegory of the marriage of Mercury and Philology, where he first coined the generic term "mechanical arts" to signify all of the useful arts and crafts—the totality of technology—Erigena not only elevates them to the "celestial" level of the liberal arts but assigns them exclusively to Mercury. If Erigena was also the first Christian to identify the useful arts as a means of restoring Adamic perfection, which accounted for their elevation, he understood that such a recovery, in overcoming "the sin of the first man," was restricted to men, that paradise would be a world without women. "At the Resurrection," he proclaimed, "sex will be abolished and nature made one. There will then be only man, as if he had never sinned." With Christ's return, as Georges Duby explained the full meaning of Erigena's words, "the end of the world would do away with dual sexuality or, more precisely, with the female part of it. When the heavens opened in glory, femininity, that imperfection, that stain on the purity of creation, would be no more."[10]

It was among the celibate monks themselves, elite men who had isolated themselves from women and assumed the burdens of female labor, that this ideological transformation of the useful arts proved most influential. For the Benedictines, especially the Cistercians, the spiritual elevation and masculinization of the arts defined their life, transforming what had heretofore been the most worldly of human activities into an other-worldly obsession. In pursuit of perfection, they mechanized myriad crafts by substituting water-power for woman-power, and thereby launched an industrial revolution of the Middle Ages. In their earthly masculine milieu, they aspired toward another, turning their heavenly attention to the useful arts, as Hugh of St. Victor indicated, "to restore within us the divine likeness."[11]

It was a rigorously reform-minded member of the Cistercian order, an austerely ascetic male enclave which strictly forbade any woman ever to cross their threshold, who gave millenarian significance and hence historical meaning and momentum to this practical project of salvation. Guided by the explicit prescriptions of the early Christian celibate John of Patmos as well as by those of his own masculine cloister, Joachim of Fiore well understood that millenarian redemption was restricted to males—and only those not "defiled by women." In his tripartite millenarian scheme, the vanguard of salvation, the *viri spirituales*, were exclusively and explicitly men only (the word *viri* being unambiguously masculine). Indeed, he identified his own brethren, the Cistercians, as the agents of transition to the new age of spiritual illumination.

If the gender identity of these saints of the millennium was not already clear enough in the writings of John of Patmos and Joachim of Fiore, it became obvious in practice less than a century after Joachim's death, as various self-anointed groups attempted to assume for themselves the mantle of the new spiritual elite. Among these were the upper-class followers of Guglielma, prophetess of Milan. Inspired by Joachim and led by Manfreda and her spiritual companion Andreas Saramita, they allotted the saintly roles of the new age to women only, thereby to ensure an absolute transformation of the corrupt world. Manfreda was to be the new pope, and her cardinals would all be women. They declared that, as the Word had become incarnate in a man, Christ, so the Holy Spirit, guide of the third stage, had become incarnate in a woman, their deceased Guglielma. But for all their zeal, their efforts were stillborn. Manfreda and her companions were burned alive, together with the disinterred bones of their prophetess. And a century later, the female Joachimite Prous Boneta, who likewise believed herself to be the incarnation of the Holy Spirit and the embodiment of the third age—the female agent of redemption as Eve had been the female agent of the Fall—met the same somber fate. There was clearly no place for women in the march toward the masculine millennium.[12]

If for Joachim the new age was represented by his fellow monks, that spiritual mantle was soon after his death claimed by another cadre of like-minded celibate males, the mendicant friars. As scholars in the forefront of learning, the friars inhabited the new celibate male cloisters of the universities. This was the setting in which the Joachimite friar Roger Bacon contemplated the past and future of the arts and sciences. Predictably, he too viewed them as exclusively male activities. From biblical accounts he traced the evolution of the arts as a strictly male affair, the remnant of Adamic perfection inherited by the "Sons of Adam," and he speculated about how their further development might contribute to a full recovery of mankind's divine birthright in a masculine millennium.[13]

As missionaries too the friars roamed the world spreading their message of salvation, as well as knowledge of the arts, and all the while maintained their distance from women. In this they were joined by the great explorers themselves—epitomized by the inspired Columbus—whose voyages excluded women. As these intrepid Westerners extended their horizons through global travel and conquest, they measured the worth of the people they encountered in exclusively male terms, not only by religious but also by technological standards. For four centuries, from 1500 to 1900, Michael Adas observed, these Westerners "assumed that the unprecedented achievements in experiment and invention which they invoked to demonstrate Western superiority"—as well as the native knowledge and tools with which these were compared—"were the products of male ingenuity and male artifice" alone.[14]

In the same spirit, Renaissance advocates of the useful arts, humanists and magi alike, pursued their antiquarian and esoteric studies in an elite male subculture and assumed that only men could hope to recapture the divine illumination they promised. Thus Marsilio Ficino and Pico della Mirandola, whose labors unearthed the ancient hermetic lore that inspired a rebirth of hermetic investigations and astrological imaginings, revived as well the ancient homoerotic ideal of intellectual purity and fraternity that came to define humanist scholarship, an impulse that meshed well with the occult pursuit of Adamic innocence and perfection. Only the "purified soul" of the magus,

Agrippa argued, could hope to return to "the condition before the Fall of Adam."[15]

In the manner of Erigena, the great alchemist Paracelsus envisioned perfection in the form of a reconstructed primordial universe before the advent of sexual duality, when Eve still remained within, and merely a part of, Adam. Indeed, Paracelsus believed that he himself had attained this primal sexual reunification within his own person, which accounted for his untroubled "natural celibacy." For Paracelsus, such purity was a precondition of the pursuit of perfection, in emulation and anticipation of its promised end. Thus, although he acknowledged that he had learned some of his knowledge of healing from wise women, he called only men to alchemical study. "Blessed be those men whose reason will reveal itself," he wrote, expressly excluding women from the art of perfection and the perfection of art. Likewise, his apocalyptically minded contemporary Albrecht Dürer addressed his inspirational instructional primer on the arts: "to our German young men I appeal alone." In his famous illustration of the Rapture from Revelation, his first great work, as the lamb appears on Mount Sion, men only await their saintly ascent.[16]

If the Reformation rekindled millenarian hopes as never before, and in the process heightened expectations of a restoration of Adamic dominion, it also aroused a resurgence of related misogynist sentiment. The same early-modern moment that spawned the intellectual ferment of the scientific revolution was also the "burning times" when countless women were persecuted as witches and perished at the stake. From the time of Luther, Steven Ozment observed, "women and marriage were widely ridiculed" and, in particular, "the biblical stories of the downfall of Adam, Samson, and David at the hands of women had gained popularity." "Oh, why did the Creator wise, /that people'd highest Heaven with spirits masculine," lamented Milton in his *Paradise Lost*, "create at last this Noveltie on Earth, this fair defect of Nature, /and not fill the World at once with men as Angels without Feminine, /or find some other way to generate Mankind?" There were no women in his *Paradise Regained*.[17]

Inspired by the Reformation, the Rosicrucians proclaimed a glorious new age of redemption through the advancement of knowledge but, like the monks and friars before them, excluded women from their blessed brotherhood. Their manifestos heralded the arrival only of "men"—not women—"embued with great wisdom, who might renew all arts and reduce them all to perfection," and thereby restore the "truth, light, life, and glory" that "the first man Adam had, which he lost in Paradise." As in all such masculine invocations of original perfection, Eve has vanished.[18]

The torch of the Rosicrucian enlightenment, and with it the pursuit of the masculine millennium, was confidently carried forth by Francis Bacon. Bacon also believed that the recovery of perfection through the arts and sciences was an exclusively male affair. Like the earlier Bacon, he assumed from biblical accounts that only men had contributed to the historical evolution of the useful arts, and also that only in chastity—"washed and clean"—would they be able to bring about their full recovery. Bacon first wrote of such a restoration in a fragment that early anticipated his magnum opus, *The Great Instauration*. Subtitled "The Great Restoration of the Power of Man over the Universe," its title, "The Masculine Birth of Time," heralded the advent of the masculine millennium. Interestingly, this provocative work, which the Bacon scholar Benjamin Farrington considered "the strongest, and from a personal angle, one of the most illuminating of all his works," was written just at the moment when the misogynist James I, whose patronage Bacon sought, succeeded Elizabeth I, who had ignored Bacon's reform proposals. Written in an avuncular style, the early essay is addressed throughout to "my son," and prescribes the means by which "to create a blessed race of Heroes and Supermen" able to "stretch the deplorable narrow limits of man's dominion over the universe to their promised bounds." "Take heart, then, my son, and give yourself to me so that I may restore you to yourself," wrote Bacon to the sons of Adam, teaching them how they might regain their rightful reign over nature and recover their prelapsarian powers. Bacon's technological utopia, *The New Atlantis,* one of his latest works, displays the same overtly masculine spirit. No women disturb the serene scientific sanctity of Solomon's House.[19]

The seventeenth-century savants who reverently followed

Bacon's lead shared the same masculine millenarian mentality. If utopian educational reformers like Comenius and Hartlib allowed that women should have access to some forms of advanced education, "it was understood," as Frank Manuel noted, "that as a rule they would be excluded from exalted studies." Robert Boyle, the virtuoso who most inspired the generation that founded the Royal Society, was a model saint as well as a model scientist and early committed himself to the celibate life. As a practitioner of the arts, he resolved in his investigations to overcome the "feminine squeamishness" that he assumed had heretofore handicapped inquiry. In the same spirit, the founding fathers of the Royal Society emphasized the quintessentially masculine nature of their enterprise. Henry Oldenburg, the society's secretary, declared that its aim was "to raise a Masculine Philosophy"; Thomas Sprat, the society's historian and chief propagandist, dubbed its domain "the Masculine Arts of Knowledge."[20]

. From the perspective of this masculine enclave, women were viewed as a threat to the entire enterprise. Walter Charleton, an early advocate of the mechanistic philosophy and a founding member of the Royal Society, gave voice to the primitive anxieties of the new men of science. "When folly hath brought us within your reach," he wrote of women, "you leap upon us and devour us. You are the traitors to Wisdom, the impediment to Industry, the clogs to virtue, and goads that drive us all to Vice, Impiety, and ruine. You are the Fools Paradise, the Wiseman's Plague, and the grand Error of Nature." Joseph Glanvill, another leading society founder and propagandist, who outlined the Baconian enterprise of Adamic restoration in his treatise on the vanity of dogmatizing, likewise cautioned the "sons of Adam" that "the Woman in us still prosecutes a deceit, like that begun in the Garden," and that their most earnest efforts were for naught so long as "our understandings are wedded to an Eve, as fatal as the mother of our Miseries." With his fellow Royal Society founder Henry More, mentor of Isaac Newton, Glanvill vigorously insisted on the existence of witches, and thereby supported the persecution of women, who were often themselves lay practitioners of the useful and healing arts. Newton himself, meanwhile, another celibate, steadfastly avoided any contact with women as he piously strove, through the study of nature and prophecy, to become one of those he called the "sons of the Resurrection."[21]

The religion of technology and its corollary myth of the masculine millennium were carried into the eighteenth century by the Freemasons, a fraternity that excluded women with a vigor worthy of monks. In its review of the history of the arts, the Freemason *Constitutions* referred only to the contributions made by men. Indeed, the opening sentence ascribed to mankind as well as to the arts an exclusively male parentage, describing Adam alone as "our first parent" in the singular, "created in the image of God," as if Eve had played no role whatever in the story of creation. Later it acidly assailed Elizabeth I for discouraging the development of the Art "because, being a WOMAN, she could not be made a Mason." Although Masonic lodges, like Cistercian monasteries, were called "mothers" and "sisters," the *Constitutions* explicitly and repeatedly excluded women from membership. Masonic practice went further, again in imitation of the Cistercians, preventing women from ever crossing the thresholds of these sacred male redoubts, which were predicated upon male-bonding rites of resurrection.[22]

A few women were briefly admitted to some lodges in revolutionary France, as Margaret Jacob has shown, but these were rare and officially condemned exceptions to the rule according to which women were excluded as "profane." A French Masonic almanac strongly advised "banishing in our assemblies the Sexe Enchanteur," and a proposal on admitting women evoked a Masonic outburst decrying women as "a vain sex, indiscreet and fickle . . . possessed of dangerous instincts. . . . We know women, their foolish spirit, their inconsequential heart. . . . Inconstancy is her only element." (This negative image of woman was represented by the Queen of the Night, who made war against the Solomonic Sarastro in Mozart's Masonic opera *The Magic Flute*.)[23]

In an initiation ritual in a Masonic lodge in Amsterdam, the initiate was asked, "In what place was the first Lodge formed?" To this he was instructed to reply, in true monastic fashion, "Upon a mountain inaccessible to the profane, where a Cock was never heard to crow, a lion to roar, or a woman to babble." When the wife of the Spanish ambassador managed to pay a brief visit to another Amsterdam lodge, as one member later recounted, the brothers were instructed, in the same monastic spirit, steadfastly to avoid the contagion. "Finally she was permitted. But before she entered, the Grand Masters asked us to cover ourselves by putting on our hats; not to look at the lady, in order to signal our disdain at all that is profane. And she entered and exited without anyone having looked at her or having given any attention to her." In the words of one late-eighteenth-century Freemason from Exeter, women were banned "because their presence might insensibly alter the purity of our maxims." "Only the friendship of men," declared the members of a new French lodge in 1761, "could produce the harmony sought in masonic society."[24]

Thus, in Freemasonry too the mythology of the masculine millennium inspired and defined the technological imagination, which was now to become incarnate, largely through Masonic agency, in engineering. By the end of the eighteenth century, Nicholas Hans observed, it was assumed that "any sound knowledge of 'useful arts and sciences' was definitely intended for boys only," and engineering was from the outset a decidedly male occupation. No doubt the formative military influence on engineering contributed to its overtly masculine character, as did the determinedly masculine mentality of the men of science who shared in its parentage and monopolized the institutions of higher learning. But ideology, specifically the religion of technology, shaped it too. As the personification of the Baconian union of science and the useful arts and the embodiment of the religion of technology that inspired it, the engineers epitomized the mythology of the masculine millennium. Held together by their own male-bonding rituals of initiation, which they inherited from the Masons, they displayed a vigorous and vigilant disdain for women and the feminine, and kept their distance lest they too, as the new Adam, forfeit their God-given powers.[25]

The culture of engineering has remained emphatically male-centered. "Engineering contains the smallest proportion of females of all major professions," sociologist Sally Hacker wrote, "and projects a heavily masculine image hostile to women." In her extended studies of the collective psychology of engineers, Hacker found that as a group they shared a starkly stratified Cartesian outlook, devaluing the body and the earth (identified with the feminine) in favor of the mind, the abstract, the mathematical. Through artifice, she suggested—a "second nature" fashioned in their own image—they sought to compensate for their social, sexual, and procreative anxieties, secure their command over the earth, and confirm their unrivaled centrality in creation.[26]

Auguste Comte, their true herald, identified the engineers as the magi of modern industry, destined to restore mankind's dominion over nature and regain the presumed primal male monopoly over the arts. Although Comte displayed a sentimental reverence for women and, late in his life, based his new religion upon the bizarre worship of his dear departed Clothilde, he nevertheless firmly believed that women were inferior beings incapable of either industrial leadership or scientific thought. In identifying women as the wellspring of love and compassion, he consigned them solely to the domestic sphere, and emphatically disqualified them from participation in the advance of modern industry, of which engineering was emblematic. Indeed, he confidently assumed, as he wrote to John Stuart Mill, that "the natural movement of our industry certainly tends gradually to pass to men the professions long exercised by women."[27]

It was at this moment that the term "technology" came into use to describe the realm of the useful arts, reshaped by science, and from the start the idea of technology became the modern measure of elite masculine identity. In the exaggeratedly masculine image of engineering especially, technological development assumed its modern appearance as a "traditionally masculine" enterprise—a mythic male affair against which women would forever have to struggle to reassert even a semblance of their former role in the useful arts. Since technology was defined from the outset as masculine, rooted as it was in the religion of technology and, hence, in the myth of a masculine millennium, women were, by definition, excluded, and whatever women did was, by definition, not included. Thus emerged what Autumn Stanley called "the stereotypes separating woman and technology," which legitimized the displacement of women, rendered their continuing contributions all but invisible, and left an indelible masculine imprint on the hallmark technological achievements of the age.

When William Broad visited the high-technology, high-security compound of the "Star Warriors" at the Lawrence Livermore Laboratories, he found that "there were no women. . . . The offices and hallways were alive with young men [but] women were nowhere to be seen." ("Just like the engineering or physics departments of any major university in America," he added.) Robert Jay Lifton noted that there was a hierarchy of esteem as well as power in this community, with the highest regard accorded the "'sons' and 'grandsons'" of Edward Teller, direct descendants of the patriarch of doom. In her study of the "nuclear language" of the "defense intellectuals," Carol Cohn described the vivid vocabulary of male competition and sexual domination that routinely came into play in discussions of nuclear warfare, and the overt phallic imagery of missiles. She and others, particularly Brian Easlea, also noted the recurring pseudo-maternal metaphors used to describe the development and detonation of atomic and hydrogen bombs from the beginning, a rhetorical masculine appropriation of the female powers of procreation which is a telltale reflection of a womanless world.[28]

If the engineers of Armageddon described the creation and "delivery" of their weapons as births, Wernher von Braun described at least one birth, that of his secretary's child, as a "successful blastoff." His world too, the enchanted enclave of space enthusiasts, was a preponderantly male domain equally marked by imagery of exclusively patriarchal procreation. This was not merely an artifact of its military origins. At the height of NASA activities, in the 1960s and 1970s, women constituted only between 2 and 3 percent of the scientific and engineering workforce (and 92 percent of clerical staff). Ian Mitroff observed an ethos of "intense masculinity" that characterized the culture of the Apollo Project. Until the space-shuttle program, all of the astronauts were men. A study on the social and psychological implications of the space program done for NASA by the Brookings Institution noted that the risk-defying "macho" astronauts were "not models for other women's husbands," and that "part of the feeling about space, which spreads right throughout the country, is women's objection to men's going there."[29]

The thoroughly masculine milieu and spirit of the space program faithfully brought to life the fantasies of its foremost visionary and inspiration, Jules Verne, "a man whom his family biographers call a misogynist." Throughout his life, Verne betrayed "a bitterness about women." Early in his career, he was a member of an elite literary dining club which called itself "onze sans Femmes," eleven without women, and throughout his long married life, he kept his distance from both his wife and his child. Verne viewed the world of science especially as an exclusively male endeavor. In a speech at a girls' school late in his life, he warned his female audience to steer clear of science and concentrate instead upon their domestic duties and destiny. "Little girls and big ones, be careful not to lose your way by running after the sciences," said Verne. "Do not plunge too deeply into science, that 'sublime emptiness' . . . wherein a man may sometimes lose himself."[30]

Verne's writings resounded with the enticements of that "sublime emptiness" as well as not a little "misogynist streak." His heroes were "peripatetic voyagers" perpetually in flight from hearth and home (and women) and steadfastly in search of some supreme fulfillment. The members of his notorious Gun Club were all men, and they exuded a mentality, inhabited a milieu, and expressed themselves in metaphors (particularly the "pre-eminently male expulsive form of the cannon") that were all at once militaristic, misogynist, monkish, apocalyptic, and transcendent—the epitome of masculine millenarianism. This was most explicit in *Sans Dessus Dessous*, the sequel and fulfillment of his *From the Earth to the Moon*.[31]

In this incredible story, the Gun Club's inspired effort to correct the earth's axis by firing off an enormous cannon buried deep into the earth ends in failure because of the ill-timed act of a woman. Distracted by a call from Mrs. Scorbitt proposing marriage, the Gun Club's leader, Maston, miscalculates the measurement of the earth's circumference, thereby dooming the mission. "Thus the fault from which the savant's downfall follows can be attributed to a woman," Verne critic Andrew Martin noted. "From the beginning of the novel, woman is denounced as the antithesis of the scientist. Maston, invoking the figure of Eve, identifies woman with the earthy, the material, the sensual, whereas man, in the figure of Newton, is credited with transcendence: while the one merely eats the apple, the other derives from it the fundamental laws of nature. . . . The text can thus be read as a re-enactment of the Fall to which it alludes at its opening: the proud and celibate Vernian bachelor succumbs to feminine guile and sheer persistence. The male paradise of perfect knowledge and control over the environment is shattered by less abstract desire."[32]

In short, in his "juvenile technological utopias," Verne offered perhaps the quintessential modern evocation of the mythology of the masculine millennium, a mythology that resonated especially in the impressionable adolescent minds of earnest young men eager to prove their manhood. This most likely explains his appeal, the remarkable fact that, by their own testimony, nearly all of the pioneers of space-flight, and a good many later enthusiasts, were as youths so inspired by Verne's vision that they resolved to dedicate their lives to making it a reality.[33]

The same mentality was also abundantly evident in the masculine domain of Artificial Intelligence, where it was simply assumed that the immortal mind was male. ("There was a standard saying in our family about Newell men, and how they were somehow so much greater than the women," Allen Newell later remembered.) Steven Levy found that the reclusive world of the computer "hackers" who developed Artificial Intelligence was characterized by "an exclusively male culture." "There were women programmers and some of them were good," he noted, "but none seemed to take hacking as a holy calling the way [the men] did. Even the substantial cultural bias against women getting into serious computing does not explain the utter lack of female hackers."[34]

At least one hacker attributed their absence "to 'genetic' or 'hardware' differences," but the close-knit, male-only cliques that typically formed the core of academic Artificial Intelligence centers probably posed the real barrier, along with the obsessive masculine computer culture these reflected. "Men tend to be seduced by the technology itself," said Oliver Strimpel, executive director of the Computer Museum in Boston. "To the truly besotted, computers are a virtual religion. . . . This is not something to be trifled with by mere females, who seem to think that machines were meant to be *used* . . .

interesting and convenient on the job but not worthy of obsession."
This same culture was evident among the "postpubescent men" of cyberspace, as well as those who inhabited the rarefied realm of Artificial Life. "Many of the engineers currently debating the form and
nature of cyberspace," sociologist Allucquere Rosanne Stone noted,
"are the young turks of computer engineering, men in their late teens
and twenties. . . ." The programmers, she found, were "almost exclusively male."[35]

"While there are certainly exceptions, many of the people doing
the work of A-life simulation at the Santa Fe Institute are men, while
most of the staff supporting the bodily and worldly needs of the researchers are women," observed anthropologist Stefan Heimreich.
These men tended to hold a "spermist view of procreation," and were
partial to patrilinear lines of descent, which were evident in their simulations and nomenclature. One researcher disdainfully described the
actual woman-centered process of pregnancy and parturition as "an
implementation problem."[36]

Finally, the brave new world of genetic engineering reflected the
same masculine millenarian culture, not only in its masculine ideal of
Adamic perfection (and parallel preoccupation with artificial reproduction) but also in its own patterns of social relations. The woeful
experience of X-ray crystallographer Rosalind Franklin, who died before her crucial contribution to the deciphering of the structure of
DNA was ever acknowledged, testified to the plight of women in this
essentially masculine world. Only much later, in the epilogue of his
account of the discovery of the double helix, did James Watson pay
belated tribute to Franklin and her work (which he had used without
her knowledge or consent), "realizing years too late the struggles that
the intelligent woman faces to be accepted by a scientific world which
often regards women as mere diversions from serious thinking." But
the situation had hardly changed. Nearly all of the acknowledged pioneers of recombinant DNA technology were men, as were the foremost architects of the Human Genome Project and practitioners of
gene therapy. Such are the new spiritual men of our age, bearers of
ancient masculine millenarian dreams now about to be realized. And
with the advent of human cloning at hand, human reproduction may

at last become their own preserve, a chaste male affair, "not defiled with women." [37]

"The changes which have taken place during the last centuries and which we sum up under the compendious term 'modern civilization,'" wrote the early-twentieth-century feminist Olive Schreiner in her classic *Woman and Labour,* "have tended to rob woman, not merely in part but almost wholly, of the more valuable part of her ancient domain of production and social labour." If women still labored mightily for human survival and continued to invent useful ways to lighten the load of mankind, their efforts went unnoticed, unrewarded, and unsung. For the advance of technology was now aimed at loftier, more transcendent, goals. As Schreiner's contemporary Sherwood Anderson observed, in an insightful essay about the mystical marriage between men and machines: "In a factual age, woman will always rule. . . . But let her come over into my own male world, the world of fancy, and surely I will lose her there." [38]

NOTES

INTRODUCTION: TECHNOLOGY AND RELIGION

1. See Mary Midgley, *Science as Salvation* (London: Routledge, 1992), and Margaret Wertheim, *Pythagoras' Trousers* (New York: Times Books, 1995).

2. Perry Miller, *The Life of the Mind in America* (New York: Harcourt, Brace and World, 1960), p. 274.

CHAPTER 1. THE DIVINE LIKENESS

1. Lynn White, Jr., "Cultural Climates and Technological Advance in the Middle Ages," *Viator*, vol. 2 (1971), pp. 172–73.

2. Norman Cohn, *The Pursuit of the Millennium* (Oxford: Oxford University Press, 1961), pp. 19, 22, 125.

3. Max Weber, *The Sociology of Religion* (Boston: Beacon Press, 1963), pp. 138, 185.

4. Gerhart B. Ladner, *The Idea of Reform* (New York: Harper and Row, 1967), pp. 91, 163, 69, 32.

5. St. Augustine, *The City of God* (Garden City, N.Y.: Doubleday, 1958), pp. 526, 527.

6. Ibid., p. 530; Jacques Ellul, "Technique and the Opening Chapters of Genesis," in Carl Mitchum and Jim Grote, eds., *Theology and Technology* (Lanham, Md.: University Press of America, 1984), p. 135.

7. White, "Cultural Climates," pp. 198–200.

8. Ibid., p. 198.

9. Ibid., p. 198; George Ovitt, *The Restoration of Perfection* (New Brunswick, N.J.: Rutgers University Press, 1986), p. 106; Ernst Benz, *Evolution and Christian Hope* (Garden City, N.Y.: Doubleday, 1975), p. 128.

10. White, "Cultural Climates," p. 198; see also David F. Noble, *A World Without Women, The Christian Clerical Culture of Western Science* (New York: Alfred A. Knopf, 1992), chap. 4.

11. Ladner, *The Idea of Reform*, pp. 2, 3.

12. Benz, *Evolution*, pp. 123–25; see also Gerhart B. Ladner, *Ad Imaginem Dei: The Image of Man in Medieval Art* (Latrobe, Pa.: Arch Abbey Press, 1965), pp. 32–34, 55.

13. Elspeth Whitney, *Paradise Restored: The Mechanical Arts from Antiquity Through the Thirteenth Century* (Philadelphia: American Philosophical Society, 1990), pp. 69, 18, 70, 71, 72, 76, 101; Ovitt, *Restoration*, p. 112; Peter Sternagel, *Die Artes Mechanicae im Mittelalter*, quoted in Whitney, *Paradise*, p. 18.

14. Martianus Capella, "The Marriage of Philology and Mercury," in William Harris Stahl and Richard Johnson, eds., *Martianus Capella and the Seven Liberal Arts* (New York: Columbia University Press, 1977), vol. 2, p. 346.

15. John J. Contreni, "John Scotus, Martin Hiberniensis: The Liberal Arts and Teaching," in Michael W. Herren, ed., *Insular Latin Studies* (Toronto: Pontifical Institute of Medieval Studies), vol. 1, p. 25.

16. Ibid., p. 26; Whitney, *Paradise*, pp. 70–72.

17. Whitney, *Paradise*, p. 72; White, "Cultural Climates," pp. 189, 197.

18. Marie-Dominique Chenu, *Nature, Man, and Society* (Chicago: University of Chicago Press, 1968), p. 43; see also Jean Gimpel, *The Medieval Machine: The Industrial Revolution of the Middle Ages* (London: Penguin, 1977).

19. White, "Cultural Climates," pp. 194–95; Cyril Stanley Smith, quoted in Lynn White, *Medieval Religion and Technology* (Berkeley: University of California Press, 1978), p. 322; Jacques Le Goff, *Time, Work, and Culture in the Middle Ages* (Chicago: University of Chicago Press, 1980), p. 78; Ovitt, *Restoration*, p. 171.

20. White, "Cultural Climates," p. 195; Whitney, *Paradise*, pp. 78, 72, 90.

21. Whitney, *Paradise*, pp. 72, 93, 90, 81.

22. Ibid., p. 81; Hugh of St. Victor, quoted in Ovitt, *Restoration*, p. 120.

23. Whitney, *Paradise*, p. 76; Ovitt, *Restoration*, pp. 121, 127; White, "Cultural Climates," p. 195.

CHAPTER 2. MILLENNIUM: THE PROMISE OF PERFECTION

1. Reinhart Maurer, "The Origins of Modern Technology in Millenarianism," in Paul T. Durbin and Friedrich Rapp, eds., *Philosophy and Technology* (Dordrecht: D. Reidel, 1983), pp. 253–65.

2. Book of Revelation, New Testament, King James Version.

3. Norman Cohn, *The Pursuit of the Millennium* (Oxford: Oxford University Press, 1961), passim.

4. See Bernard McGinn, *The Calabrian Abbot: Joachim of Fiore in the History of Western Thought* (New York: Macmillan, 1985); Bernard McGinn, "Apocalyptic Traditions and Spiritual Identity in Thirteenth Century Religious Life," in E. Rozanne Elder, ed., *The Roots of the Modern Christian Tradition* (Kalamazoo, Mich.: Cistercian Publications, 1984).

5. Ernst Benz, *Evolution and Christian Hope* (Garden City, N.Y.: Doubleday, 1975), p. 36.

6. Frank E. Manuel, *Freedom from History* (New York: New York University Press, 1971), p. 127; see also Marjorie Reeves, *The Influence of Prophecy in the Later Middle Ages: A Study in Joachimism* (Oxford: Oxford University Press, 1969); Richard K. Emerson and Bernard McGinn, eds., *The Apocalypse in the Middle Ages* (Ithaca, N.Y.: Cornell University Press, 1992).

7. Cohn, *Pursuit of the Millennium*, pp. 108–11.

8. Arnold Pacey, *The Maze of Ingenuity: Ideas and Idealism in the Development of Technology* (Cambridge, Mass.: MIT Press, 1976), p. 58.

9. Will Durant, *The Age of Faith* (New York: Simon and Schuster, 1950), p. 1010; J. B. Bury, *The Idea of Progress* (London: Macmillan, 1928), p. 26.

10. Roger Bacon, *The Opus Majus of Roger Bacon* (New York: Russell and Russell, 1962), p. 417 and passim; Stewart C. Easton, *Roger Bacon and His Search for a Universal Science* (New York: Russell and Russell, 1971), passim; Pacey, *Maze of Ingenuity*, pp. 56–57.

11. Bacon, *Opus Majus*, pp. 633–34.

12. Ibid., pp. 52, 65.

13. Bacon, quoted in J. B. Bury, *The Idea of Progress* (London: Macmillan, 1928), p. 26.

14. Lynn Thorndike, *History of Magic and Experimental Science* (New York: Columbia University Press, 1934), vol. 2, pp. 863–65, 842.

15. Ibid., vol. 3, pp. 347–55; Robert P. Multhauf, "John of Rupescissa and the Origins of Medical Chemistry," *Isis*, vol. 45 (1954), pp. 359–66.

16. John Leddy Phelan, *The Millennial Kingdom of the Franciscans in the New World* (Berkeley: University of California Press, 1970), p. 1.

17. Pauline Moffitt Watts, "Prophecy and Discovery: On the Spiritual Origins of Christopher Columbus' Enterprise of the Indies," *American Historical Review*, vol. 90 (1985), pp. 73–102.

18. Leonard I. Sweet, "Christopher Columbus and the Millennial Vision of the New World," *Catholic Historical Review*, vol. 72 (July 1986), pp. 369–82; Thorndike, *History of Magic*, vol. 3, p. 842.

19. Watts, "Prophecy and Discovery," passim.

20. Ibid.

21. Phelan, *Millennial Kingdom*, passim.

22. Ibid.

23. Ibid.

24. Thorndike, *History of Magic*, vol. 4, p. 107.

25. Kirkpatrick Sale, *The Conquest of Paradise* (New York: Alfred A. Knopf, 1992), p. 190; Watts, "Prophecy and Discovery," p. 73.

26. Sale, *The Conquest of Paradise*, pp. 188, 190.

27. Ibid., p. 175.

CHAPTER 3. VISIONS OF PARADISE

1. Marjorie Reeves, *The Influence of Prophecy in the Later Middle Ages: A Study in Joachimism* (Oxford: Oxford University Press, 1969), pp. 174, 268, 431, 438.

2. Charles G. Nauert, Jr., *Agrippa and the Crisis of Renaissance Thought* (Urbana: University of Illinois Press, 1965), pp. 48, 49, 284; see also Frances Yates, *The Rosicrucian Enlightenment* (Boulder: Shambala Press, 1978), p. 119, and Reeves, *Prophecy*, p. 102.

3. Reeves, *Prophecy*, p. 454; P. M. Rattansi, "The Social Interpretation of Science in the Seventeenth Century," in Peter Mathias, ed., *Science and Society, 1600–1900* (Cambridge: Cambridge University Press, 1972), p. 11; Jolande Jacobi, ed., *Paracelsus, Selected Writings* (New York: Pantheon, 1951), pp. 201, 257.

4. Ibid., p. 296.

5. Wilhelm Waetzgoldt, *Dürer and His Time* (London: Phaidon Press, 1950), pp. 15, 32.

6. John Leddy Phelan, *The Millennial Kingdom of the Franciscans in the New World* (Berkeley: University of California Press, 1970), pp. 70–72; Frank E. Manuel, *Freedom from History* (New York: New York University Press, 1971), p. 91; Ernest Lee Tuveson, *Millennium and Utopia* (New York: Harper and Row, 1964), pp. 22–30; Katherine R. Firth, *The Apocalyptic Tradition: Reformation Britain, 1530–1645* (New York: Oxford University Press, 1979), p. 248.

7. Giordano Bruno, "The Expulsion of the Triumphant Beast," quoted in Benjamin Farrington, *The Philosophy of Francis Bacon* (Chicago: University of Chicago Press, 1964), p. 27.

8. George Ovitt, "Critical Assessments of Technology from Campanella to the Harringtonians," unpublished manuscript, 1989.

9. Manuel, *Freedom from History*, p. 91: Rattansi, "Social Interpretation," p. 12.

10. Yates, *Rosicrucian Enlightenment*, passim.

11. Ibid.

12. Ibid.

13. Ibid.

CHAPTER 4. PARADISE RESTORED

1. Katherine R. Firth, *The Apocalyptic Tradition: Reformation Britain, 1530–1645* (New York: Oxford University Press, 1979), p. 3.

2. Henry Guppy, *William Tindale and the Earlier Translations of the Bible into English* (Manchester: University Press, 1925), pp. 28–29.

3. Gustavus S. Paine, *The Man Behind the King James Version* (Grand Rapids, Mich.: Baker Book House, 1959).

4. Trevelyan quoted in Guppy, *William Tindale*, p. 29; Christopher Hill, *The English Bible and the Seventeenth Century Revolution* (Allen Lane: Penguin Press, 1993), pp. 27, 34.

5. Keith Thomas, *Man and the Natural World* (New York: Pantheon, 1983), pp. 18, 22.

6. Arnold Williams, *The Common Expositor: An Account of the Commentaries on Genesis, 1527–1633* (Chapel Hill: University of North Carolina Press, 1948).

7. Charles Webster, *Great Instauration: Science, Medicine, and Reform 1626–1660* (London: Gerald Duckworth, 1975), pp. 324, 326, 328; Hill, *English Bible*, p. 34.

8. Firth, *Apocalyptic Tradition*, p. 206; Richard H. Popkin, ed., *Millenarianism and Messianism in English Literature and Thought, 1650–1800* (Leiden: E. J. Brill, 1988), p. 5; Bernard Capp, *Astrology and the Popular Press: English Almanacs, 1500–1800* (London: Faber and Faber, 1979), p. 169.

9. Charles Whitney, *Francis Bacon and Modernity* (New Haven: Yale University Press, 1986), pp. 44, 45.

10. William M. Lamont, *Godly Rule: Politics and Religion, 1603–1660* (New York: St. Martin's Press, 1969), pp. 31, 13; Hill, *English Bible*, p. 304.

11. Popkin, ed., *Millenarianism and Messianism*, pp. 6, 7; Milton quoted in Webster, *Great Instauration*, p. 1.

12. Ernest Lee Tuveson, *Millennium and Utopia* (New York: Harper and Row, 1964), p. 84; Webster, *Great Instauration*, p. 18, 509, 335 (Milton quote).

13. Ibid., p. 335; P. M. Rattansi, "The Social Interpretation of Science in the Seventeenth Century," in Peter Mathias, ed., *Science and Society, 1600–1900* (Cambridge: Cambridge University Press, 1972), p. 13.

14. Lewis Mumford, *Pentagon of Power* (New York: Harcourt Brace Jovanovich, 1964), p. 106; Bacon quoted in Robert Merton, *Science, Technology, and Society in Seventeenth Century England* (New York: Howard Fertig, 1970), p. 115; see also Margaret C. Jacob, *The Cultural Meaning of the Scientific Revolution* (Philadelphia: Temple University Press, 1988), pp. 32, 35; James R. Jacob, "By an Orphean Charm," in Phyllis Meade and Margaret C. Jacob, eds., *Politics and Culture in Early Modern Europe* (Cambridge: Cambridge University Press, 1988), p. 236.

15. Webster, *Great Instauration*, pp. 336, 335; George Ovitt, *The Restoration of Perfection* (New Brunswick, N.J.: Rutgers University Press, 1986), p. 17; Francis Bacon, *Novum Organum,* in Benjamin Farrington, ed., *The Works of Francis Bacon* (Philadelphia: Carey and Hart, 1848), vol. 4, p. 247.

16. Paolo Rossi, *Francis Bacon: From Magic to Science* (London: Routledge and Kegan Paul, 1968), pp. 7–11.

17. Jacob, *Cultural Meaning*, p. 32; Mumford, *Pentagon of Power*, p. 106; Francis Bacon, "The Masculine Birth of Time," in Benjamin Farrington, *The Philosophy of Francis Bacon* (Chicago: University of Chicago Press, 1964), p. 72; Bacon, *Novum Organum,* aphorism 68.

18. Rossi, *Francis Bacon*, pp. 127–29; Frances Yates, *The Rosicrucian Enlightenment* (Boulder: Shambala Press, 1978), p. 119.

19. Yates, *Rosicrucian Enlightenment*, p. 119; Farrington, *Philosophy*, p. 21; Rossi, *Francis Bacon*, p. 127; Bacon, "Masculine Birth of Time," p. 72.

20. Francis Bacon, *Novum Organum,* in Farrington, ed., *Works*, vol. 4, pp. 247–48; Francis Bacon, *Valerius Terminus,* in Farrington, ed., *Works,* vol. 3, pp. 217, 219; vol. 4, pp. 21, 247–48; Bacon, "The Refutation of Philosophies," in Farrington, *Philosophy*, p. 106; see also Eugene Klaaren, *Religious Origins of Modern Science: Belief in Creation in Seventeenth Century Thought* (Grand Rapids, Mich.: William B. Eerdman, 1977), p. 92.

21. Thomas, *Man and the Natural World*, p. 18; Webster, *Great Instauration*, p. 329; Bacon, Preface to *The Great Instauration,* in *The Physical and Metaphysical Works of Lord Bacon* (London: George Bell and Sons, 1904), p. 9.

22. Bacon, *Valerius Terminus; Novum Organum; Great Instauration.*

23. Yates, *Rosicrucian Enlightenment*, p. 129.

24. Webster, *Great Instauration*, pp. 511, 47, 22–23; Whitney, *Francis Bacon and Modernity*, p. 44.

25. Webster, *Great Instauration*, pp. 69, 192.

26. Samuel Hartlib, "Petition to Parliament (1649)," in Webster, *Great Instauration*, appendix; Merton, *Science, Technology, and Society*, pp. 116–17; Rattansi, "Social Interpretation," p. 20.

27. Webster, *Great Instauration*, pp. 324, 326, 328, 329.

28. Ibid., p. 246, appendix.

29. Firth, *Apocalyptic Tradition*, pp. 206, 213; Milton, quoted in Webster, *Great Instauration*, p. 100.

CHAPTER 5. HEAVENLY VIRTUOSI

1. Lewis Mumford, *Pentagon of Power* (New York: Harcourt Brace Jovanovich, 1964), p. 111.

2. Robert Merton, *Science, Technology, and Society in Seventeenth Century England* (New York: Howard Fertig, 1970), p. 81; Charles Webster, *The Great Instauration: Science, Medicine, and Reform, 1626–1660* (London: Gerald Duckworth and Co., 1975, p. 67; Robert Boyle, "Of the Usefulness of Natural Philosophy," in *Works of the Honorable Robert Boyle* (London, 1772), vol. 2, p. 5.

3. Webster, *Great Instauration*, pp. 99, 162, 496.

4. Merton, *Science, Technology, and Society*, appendix; Mumford, *Pentagon of Power*, p. 116.

5. Margaret C. Jacob, *The Cultural Meaning of the Scientific Revolution* (Philadelphia: Temple University Press, 1988), pp. 34, 75.

6. P. M. Rattansi, "The Social Interpretation of Science in the Seventeenth Century," in Peter Mathias, ed., *Science and Society, 1600–1900* (Cambridge: Cambridge University Press, 1972), p. 22; Boyle, quoted in Ernest Lee Tuveson, *Millennium and Utopia* (New York: Harper and Row, 1964), p. 102; Robert Boyle, "The Christian Virtuoso," in *Works of Robert Boyle*, vol. 6, pp. 776–89; Eugene M. Klaaren, *The Religious Origins of Modern Science* (Grand Rapids, Mich.: William B. Eerdman, 1977), p. 129.

7. Klaaren, *Religious Origins*, p. 129; Webster, *Great Instauration*, p. 29; Merton, *Science, Technology, and Society*, p. 104; Margaret C. Jacob, "Millenarianism and Science in the Late Seventeenth Century," *Journal of the History of Ideas*, vol. 37 (1976), p. 338.

8. Joseph Glanvill, *The Vanity of Dogmatizing* (New York: Columbia University Press, 1931), pp. 3–5, 6, 8, 11.

9. Ibid., pp. 238–41.

10. Milton, quoted in Webster, *Great Instauration*, p. 100.

11. Amos Funkenstein, *Theology and the Scientific Imagination* (Princeton: Princeton

University Press, 1986), p. 298; Joseph Glanvill, *Plus Ultra or the Progress and Advancement of Knowledge*, quoted in Funkenstein, *Theology*, p. 298; Boyle, "Usefulness," p. 7.

12. Funkenstein, *Theology*, p. 299; Klaaren, *Religious Origins*, pp. 100, 190, 128.

13. Boyle, "Usefulness," pp. 54, 32; Tuveson, *Millennium and Utopia*, p. 100; Klaaren, *Religious Origins*, p. 105.

14. Frank E. Manuel, *The Religion of Isaac Newton* (Oxford: Oxford University Press, 1974), pp. 97, 91, 99, 100, 47.

15. Rattansi, "Social Interpretation," p. 22; Jacob, "Millenarianism and Science," p. 340; Manuel, *Religion of Isaac Newton*, p. 99; see also Arthur Quinn, "On Reading Newton Apocalyptically," in Richard H. Popkin, ed., *Millenarianism and Messianism in English Literature and Thought, 1650–1800* (Leiden: E. J. Brill, 1988), pp. 176–92.

16. Klaaren, *Religious Origins*, p. 15.

17. Rattansi, "Social Interpretation," p. 21; Milton, quoted in Webster, *Great Instauration*, p. 100.

18. Klaaren, *Religious Origins*, pp. 111, 85, 93.

19. John Beale to Robert Boyle, October 17, 1663, quoted in Tuveson, *Millennium and Utopia*, p. 110; John Edwards, quoted in ibid., p. 131.

20. Francis Bacon, *The Great Instauration*, quoted in Klaaren, *Religious Origins*, p. 95; Mumford, *Pentagon of Power*, pp. 117, 125.

21. Boyle, quoted in Tuveson, *Millennium and Utopia*, p. 110; Boyle, "Usefulness," p. 14.

CHAPTER 6. THE NEW ADAM

1. Margaret C. Jacob, "Millenarianism and Science in the Late Seventeenth Century," *Journal of the History of Ideas*, vol. 37 (1976), p. 335.

2. Burnet, quoted in Ernest Lee Tuveson, *Millennium and Utopia* (New York: Harper and Row, 1964), pp. 128, 122.

3. Burnett, Lord Monboddo, quoted in ibid., p. 190.

4. Clarke Genett, "Joseph Priestley, the Millennium, and the French Revolution," *Journal of the History of Ideas*, vol. 34 (1973), p. 51.

5. Ibid., pp. 51, 61.

6. Ibid., pp. 61, 55.

7. Ibid., pp. 63.

8. Geoffrey Cantor, *Michael Faraday: Sandemanian and Scientist* (London: Macmillan, 1991), pp. 8, 292, 294.

9. Lewis Campbell and William Garnett, *The Life of James Clerk Maxwell* (New York: Johnson Reprint Corporation, 1969) (original 1882), p. 323.

10. Charles Babbage, *The Ninth Bridgewater Treatise* (London: Frank Cass, 1967) (original 1837), pp. 82, 92–93, 132, 139–40, 164, 173; Linda M. Strauss, "Automata: A Study in the Interface of Science, Technology, and Popular Culture," unpublished Ph.D. dissertation, University of California, San Diego, 1987.

11. Margaret Jacob, *Living the Enlightenment* (New York: Oxford University Press, 1991), pp. 22, 204.

12. Nicholas Hans, "UNESCO of the Eighteenth Century: La Loge des Neuf Soeurs and Its Venerable Master Benjamin Franklin," *Proceedings of the American Philosophical Society*, vol. 9, no. 5 (Oct. 1953), p. 513; Margaret C. Jacob, *The Cultural Meaning of the Scientific Revolution* (Philadelphia: Temple University Press, 1988), pp. 126–28; Frances Yates, *The Rosicrucian Enlightenment* (Boulder: Shambala Press, 1978), pp. 209–10. See also Margaret Jacob, "Freemasonry and the Utopian Impulse," in Richard H. Popkin, ed., *Millenarianism and Messianism in English Literature and Thought, 1650–1800* (Leiden: E. J. Brill, 1988); R. William Weisberger, *Speculative Freemasonry and the Enlightenment* (New York: Columbia University Press, 1993); and David Stevenson, *The Origins of Freemasonry* (Cambridge: Cambridge University Press, 1988).

13. Yates, *Rosicrucian Enlightenment*, p. 210; Nicholas Hans, *New Trends in Education in the Eighteenth Century* (London: Routledge and Kegan Paul, 1951), p. 139; Jacob, *Living the Enlightenment*, pp. 36–37.

14. Hans, *New Trends*, pp. 85, 40.

15. *The Constitutions of the Free-Masons* (New York: J. W. Leonard., 1855), p. 1.

16. Jacob, *Cultural Meaning*, pp. 157–59.

17. Yates, *Rosicrucian Enlightenment*, p. 83; Jacob, *Living the Enlightenment*, p. 208; Abner Cohen, "The Politics of Ritual Secrecy," *Man*, vol. 6 (Sept. 1977), p. 137.

18. Strauss, "Automata," pp. 41, 79.

19. John Spargo, *Freemasonry in Vermont, 1865–1944* (Burlington: Grand Lodge of Vermont, 1944), pp. 99–150; "Letters to the Editor," *Friend* (St. Johnsbury, Vt.), vol. 1, no. 1 (July 22, July 29, and Aug. 5, 1829); Jacob, "Freemasonry and the Utopian Impulse," p. 137; Jacob, *Living the Enlightenment*, p. 184.

20. Jacob, *Living the Enlightenment*, p. 208; Jacob, *Cultural Meaning*, p. 148.

21. Jacob, *Cultural Meaning*, p. 186; Hans, *New Trends*, pp. 58–59.

22. Hans, *New Trends*, pp. 213, 154; Hans, "UNESCO," p. 513.

23. Weisberger, *Speculative Freemasonry*, pp. 79–80.

24. Eric Dorn Brose, *The Politics of Technological Change in Prussia* (Princeton: Princeton University Press, 1992), chap. 6.

25. Jacob, *Cultural Meaning*, p. 157; Alexander Gibb, *The Story of Telford: The Rise of Civil Engineering* (London: Alexander Maclehouse, 1935), pp. 11, 36.

26. Frederick B. Artz, *The Development of Technical Education in France* (Cambridge, Mass.: MIT Press, 1966), p. 83; see also William E. Wickenden, "A Comparative Study of the Engineering Education in the U.S. and Europe," in *Report of the Investigation of Engineering Education* (Pittsburgh: Society for the Promotion of Engineering Education, 1930), vol. 1, pp. 807–24, and Antoine Picon, *French Architects and Engineers in the Age of Enlightenment* (Cambridge: Cambridge University Press, 1992), pp. 346–53.

27. Artz, *Development of Technical Education*, pp. 98, 101; see also Alain le Bihan, *Loges et Chapitres de la Grande Loge et du Grand Orient de France* (Paris: Bibliothèque Nationale, 1967), pp. 390, 418; John H. Weiss, *The Making of Technological Man: The Origins of French Engineering Education* (Cambridge, Mass.: MIT Press, 1982), p. 93.

28. Artz, *Development of Technical Education*, pp. 153–55, 98–101; Michelle Sadoun-Goupil, *Le Chimiste Claude-Louis Berthollet* (Paris: Librairie Philosophique J. Vrin, 1977), pp. 61–62; E. T. Bell, *Men of Mathematics* (New York: Dover Publications, 1937), pp. 183–205; Bihan, *Loges et Chapitres*, pp. 356–58; Jacob, *Living the Enlightenment*, p. 146.

29. Olivier, quoted in Weiss, *Making of Technological Man*, pp. 157–58; see also Hans, "UNESCO," p. 323; Artz, *Development of Technical Education*, p. 249.

30. Weiss, *Making of Technological Man*, pp. 157, 182.

31. W. H. G. Armytage, *The Rise of the Technocrats* (London: Routledge and Kegan Paul, 1965), pp. 66, 72; Weiss, *Making of Technological Man*, p. 94; Auguste Comte, "Third Essay," in Gertrude Lenzer, *Auguste Comte and Positivism: The Essential Writings* (New York: Harper and Row, 1975), pp. 89–90; Auguste Comte, "Fourth Essay," quoted in Lewis Mumford, *Technics and Civilization* (New York: Harcourt, Brace and World, 1934), pp. 219–20.

32. F. J. Gould, *The Life Story of Auguste Comte* (Austin: American Atheist Press, 1984), pp. 5, 29, 34; Auguste Comte, "Cours de Philosophie Positive," in Lenzer, *Comte and Positivism*, p. 81; Frank E. Manuel, *Freedom from History* (New York: New York University Press, 1971), p. 59.

33. Lenzer, *Comte and Positivism*, pp. 18, 23, xxxii; Edward Caird, *The Social Philosophy and Religion of Auguste Comte* (Glasgow: James Maclehouse and Sons, 1845), p. xv; Auguste Comte, "Système de Politique Positive," in Lenzer, *Comte and Positivism*, pp. 452, 466; Armytage, *Rise of Technocrats*, p. 72; Comte, "Cours de Philosophie Positive," p. 302.

34. Comte, "Système," pp. 466, 457, 444, 453, 447, 457, 458.

35. Comte, "Third Essay," p. 32.

36. Comte, "Système," pp. 458, 449; Comte, "Third Essay," p. 25.

37. Gould, *Life Story of Auguste Comte*, pp. 19, 59.

38. Comte, "Système," pp. 466, 447, 474; see also Max Horkheimer, *The Eclipse of Reason* (Oxford: Oxford University Press, 1947), p. 101.

39. Robert Owen, *Debate on the Evidences of Christianity* (London: R. Groombridge, 1839), pp. 28, 36; Maxine Berg, *The Machinery Question* (Cambridge: Cambridge University Press, 1980), p. 271.

40. Berg, *Machinery Question*, p. 278.

41. Armytage, *Rise of Technocrats*, p. 112.

CHAPTER 7. THE NEW EDEN

1. Allen E. Roberts, *Freemasonry in American History* (Richmond, Va.: Macoy Publishing Company, 1985), pp. 328, 340, 382, 384; Bobby J. Demott, *Freemasonry in American Culture and Society* (Washington, D.C.: University Press of America, 1980), p. 254.

2. R. W. B. Lewis, *The American Adam* (Chicago: University of Chicago Press, 1955), pp. 4, 5, 10; David W. Noble, *The Eternal Adam and the New World Garden* (New York: George Braziller, 1968), pp. 5, 36; Joel Nydahl, "Introduction," in John Adolphus Etzler, *The Collected Works of John Adolphus Etzler* (Delmar, N.Y.: Scholar's Facsimiles and Reprints, 1977), p. ix; Walt Whitman, quoted in Lewis, *The American Adam*, p. 5.

3. Paul Boyer, *When Time Shall Be No More: Prophecy Belief in Modern American Culture* (Cambridge, Mass.: Harvard University Press, 1992), p. 68.

4. Ibid., p. 71.

5. Perry Miller, *The Life of the Mind in America* (New York: Harcourt, Brace and World, 1960), pp. 7, 272; Robert Fletcher, quoted in David F. Noble, *A World Without Women* (New York: Oxford University Press, 1992), pp. 246–47.

6. Timothy P. Weber, *Living in the Shadow of the Second Coming* (New York: Oxford University Press, 1979), p. 102; Miller, *Life of the Mind*, p. 274.

7. Nydahl, "Introduction," pp. xi, xii.

8. Etzler, *Collected Works*, pp. 56, 82, 98, 117–18; Henry David Thoreau, "Paradise (to be) Regained," *United States Magazine and Democratic Review*, vol. 13 (Nov. 1843), pp. 451–63, reprinted in Thomas Parke Hughes, *Changing Attitudes Toward American Technology* (New York: Harper and Row, 1975), p. 90.

9. Patrick Brostowin, "John Adolphus Etzler: The Scientific-Utopian During the 1830's and 1840's," unpublished Ph.D. dissertation, New York University, 1969, p. 17, quoted in Nydahl, "Introduction," p. xv; Etzler, *Collected Works*, pp. 4, 49, 79.

10. Ethel M. McAllister, *Amos Eaton: Scientist and Educator* (Philadelphia: University of Pennsylvania Press, 1941), pp. 368, 491, 490.

11. Miller, *Life of the Mind*, p. 289; Jacob Bigelow, *Elements of Technology* (Boston: Boston Press, 1829), p. 4; John Beekmann, *A History of Inventions and Discoveries* (London: J. Walker, 1814), p. x; Jacob Bigelow, *Remarks on Classical and Utilitarian Studies* (to American Arts and Sciences, Dec. 20, 1866) (Boston: Little, Brown, 1867), p. 11; Jacob Bigelow, "A Poem on Professional Life: An Address to Cambridge Phi Beta Kappa," unpublished manuscript, George Ticknor Collection File #001474, Dartmouth College; Jacob Bigelow, address at MIT, 1865, quoted in Howard P. Segal, *Technological Utopianism in American Culture* (Berkeley: University of California Press, 1985), p. 81.

12. Leo Marx, *The Pilot and the Passenger* (New York: Oxford University Press, 1988), p. 5; James W. Carey, *Communication as Culture* (Boston: Unwin Hyman, 1989), p. 120; Segal, *Technological Utopianism*, p. 94; Emerson, quoted in Thomas P. Hughes, "The Second Creation of the World," unpublished manuscript, no pagination or date.

13. Carleton Mabee, *The American Leonardo: A Life of Samuel F. B. Morse* (New York: Alfred A. Knopf, 1944), pp. 260, 275, 369; *Dictionary of American Biography* (New York: American Society of Learned Societies, 1934), vol. 7, pp. 247–51.

14. Carey, *Communication as Culture*, pp. 206–7.

15. George Babcock, ASME Transactions (1888), quoted in Bruce Sinclair, "Local History and National Culture: Notes on Engineering Professionalism in America," *Technology and Culture*, vol. 17 (Oct. 1986), p. 692; Segal, *Technological Utopianism*, p. 94.

16. George S. Morison, *The New Epoch: As Developed by the Manufacture of Power* (Boston: Houghton Mifflin, 1903), pp. 4, 75.

17. Ibid., pp. 5, 6, 11, 68, 75, 128, 130, 132–33.

18. Robert Thurston, "Scientific Research: The Art of Revelation and Prophecy," *Science*, vol. 16 (Sept. 12–19, 1902), pp. 402, 404, 407, 422, 423.

19. Ibid., pp. 455, 457.

20. Ralph E. Flanders, "The New Age and the New Man," in Charles A. Beard, ed., *Toward Civilization* (New York: Longmans, Green, 1930), p. 23; Albert Merrill, *The Great Awakening*, cited in Segal, *Technological Utopianism*, p. 48; Edison quotes drawn from materials on display at Edison Home Museum, Fort Myers, Florida; James Newton, *Uncommon Friends* (New York: Harcourt Brace, 1987), pp. 29–30, 229.

21. Neil Baldwin, *Edison: Inventing the Century* (New York: Hyperion, 1995), pp. 172, 96, 375, 377.

22. Milton Cantor, "The Backward Look of Bellamy's Socialism," in Daphne Patai, ed., *Looking Backward, 1988–1888* (Amherst: University of Massachusetts Press,

1988), p. 20; Sylvia E. Bowman, *The Year 2000: A Critical Biography of Edward Bellamy* (New York: Bookman Associates, 1958), pp. 21, 36.

23. Edward Bellamy, *The Religion of Solidarity* (Folcroft, Pa.: Folcroft Press, 1940) (original 1874), pp. 16, 21, 22, 43.

24. Howard P. Segal, "Bellamy and Technology," in Patai, ed., *Looking Backward,* pp. 91, 104.

25. Edward Bellamy, *Looking Backward* (New York: New American Library, 1960), pp. 218, 185, 190, 191, 194.

26. Ibid., pp. 220, 222.

27. Edward Bellamy, *Equality* (New York: D. Appleton and Co., 1897), pp. 235, 236.

CHAPTER 8. ARMAGEDDON: ATOMIC WEAPONS

1. Perry Miller, "The End of the World," in Perry Miller, *Errand into the Wilderness* (New York: Harper and Row, 1956), p. 235.

2. Ibid., p. 235; Ernest R. Sandeen, *The Roots of Fundamentalism* (Chicago: University of Chicago Press, 1970), p. 233; Michael Sherry, *The Rise of American Air Power: The Creation of Armageddon* (New Haven: Yale University Press, 1987), p. 330; Paul Boyer, *By the Bombs' Early Light* (New York: Pantheon, 1985), p. 238.

3. Lewis Mumford, *Pentagon of Power* (New York: Harcourt Brace Jovanovich, 1964), p. 47; Richard Rhodes, *The Making of the Atomic Bomb* (New York: Simon and Schuster, 1986), pp. 21–23.

4. Leo Szilard, quoted in Rhodes, *Atomic Bomb,* p. 25.

5. Ernest Rutherford, quoted in Brian Easlea, *Fathering the Unthinkable* (London: Pluto Press, 1983), p. 43.

6. Rhodes, *Atomic Bomb,* p. 571; Robert Oppenheimer to Leslie Groves, Oct. 20, 1962, quoted in ibid., p. 572.

7. John Donne, *The Divine Poems* (Oxford: Clarendon Press, 1952), p. 50.

8. Rhodes, *Atomic Bomb,* p. 572; unidentified quotation from film *The Day After Trinity* (Pyramid Films, 1980), quoted in Sally M. Gearhart, "An End to Technology," in Joan Rothschild, ed., *Machina ex Dea* (New York: Pergamon Press, 1983), p. 177.

9. Oppenheimer, quoted in Rhodes, *Atomic Bomb,* p. 676; George Kistiakowski, quoted in Robert Jay Lifton and Eric Markusen, *The Genocidal Mentality* (New York: Basic Books, 1988), p. 83; Farrell, quoted in William Laurence, *Dawn over Zero* (London: Museum Press, 1974), pp. 198–99.

10. Miller, "End of the World," pp. 238, 219, 235, 238; Carol Cohn, "Nuclear Language," *Bulletin of the Atomic Scientists*, June 1987, p. 70.

11. Boyer, *Bombs' Early Light*, p. 237; Fallow and Tittle, quoted in ibid., pp. 237–38; Churchill, quoted in Easlea, *Fathering*, p. 103; Hutchins, quoted in Sherry, *American Air Power*, p. 353.

12. Billy Graham, quoted in Grace Halsell, *Prophecy and Politics: Militant Evangelists and the Road to Nuclear War* (Westport, Conn.: Lawrence Hill, 1986), p. 28.

13. Jerry Falwell, quoted in ibid., pp. 34, 39.

14. A. G. Mojtabai, *Blessed Assurance* (Boston: Houghton Mifflin, 1986), pp. 78, 80, 84.

15. Ibid., pp. 152, 157, 167.

16. Lifton and Markusen, *Genocidal Mentality*, pp. 112, 85.

17. Ibid., pp. 86–87, 88.

18. Ibid., pp. 83, 141; Herbert York, quoted in William J. Broad, *Star Warriors: A Penetrating Look into the Lives of the Young Scientists Behind Our Space Age Weaponry* (New York: Simon and Schuster, 1985), p. 217.

19. Lifton and Markusen, *Genocidal Mentality*, p. 118; Broad, *Star Warriors*, pp. 190, 65, 173.

20. Broad, *Star Warriors*, pp. 65, 127.

21. Freeman Dyson, "Human Consequences of the Exploration of Space," *Bulletin of the Atomic Scientists*, Sept. 1969; Freeman Dyson, "Space Traveler's Manifesto," cited in Mary Midgley, *Science as Salvation* (London: Routledge, 1992), p. 184; Freeman Dyson, "Time Without End," *Review of Modern Physics*, vol. 51, no. 3 (July 1979); Louis J. Halle, "A Hopeful Future for Mankind," *Foreign Affairs*, Summer 1980; Easlea, *Fathering*, p. 147; Rhodes, *Atomic Bomb*, p. 25; Broad, *Star Warriors*, pp. 127, 131; Miller, "End of the World," p. 235.

CHAPTER 9. THE ASCENT OF THE SAINTS:
SPACE EXPLORATION

1. Marjorie Hope Nicolson, *Voyages to the Moon* (New York: Macmillan, 1948), pp. 20, 27; see also John Wilkins, *The Discovery of a World in the Moone* (Delmar, N.Y.: Scholar's Facsimiles and Reprints, 1973), p. 191.

2. Carola Baumgardt, *Johannes Kepler: Life and Letters* (New York: Philosophical Library, 1951), pp. 23, 31, 44, 114, 34; John Lear, *Kepler's Dream* (Berkeley: University of California Press, 1965), p. 76.

3. Baumgardt, *Johannes Kepler*, pp. 155, 175; Edward Rosen, trans., *Kepler's "Somnium"* (Madison: University of Wisconsin Press, 1967), p. 33; Arthur Koestler, *The Sleepwalkers* (New York: Pelican, 1959), p. 378.

4. Nicolson, *Voyages,* pp. 40, 47; John Wilkins, *A Discourse Concerning a New World and Another Planet* (Delmar, N.Y.: Scholar's Facsimiles and Reprints, 1973), pp. 241–42, 243; Wilkins, *Discovery of a World,* pp. 205, 208.

5. Nicolson, *Voyages,* pp. 123, 59, 60; Peter Costello, *Jules Verne, Inventor of Science Fiction* (London: Hodder and Stoughton, 1978), p. 36.

6. Jules Verne, *From the Earth to the Moon* (New York: Dodd, Mead, 1962), p. 3.

7. Ibid., pp. 89, 211, 284.

8. Andrew Martin, *The Mask of the Prophet: The Extraordinary Fictions of Jules Verne* (New York: Oxford University Press, 1990), pp. 187, 188.

9. Ibid., pp. 192–93.

10. Ray Bradbury, "Foreword," in William Butcher, *Verne's Journey to the Centre of the Self* (New York: Macmillan, 1990), pp. xiii, xiv, xv. See also Andrew Martin, *The Knowledge of Ignorance: From Genesis to Jules Verne* (Cambridge: Cambridge University Press, 1985); Jean Chesneaux, *The Political and Social Ideas of Jules Verne* (London: Thames and Hudson, 1972); Lear, *Kepler's Dream,* p. 76.

11. Michael Sherry, *The Rise of American Air Power: The Creation of Armageddon* (New Haven: Yale University Press, 1987), pp. 209–10.

12. Tom Crouch, *The Bishop's Boys* (New York: W. W. Norton, 1989), p. 33; "Thanked for Not Flying," *New York Times,* November 4, 1910, p. 2.

13. K. E. Tsiolkovsky, "Autobiography," in Arthur C. Clarke, *The Coming of the Space Age* (New York: Meredith Press, 1967), pp. 100, 101, 104; Walter A. McDougall, *The Heavens and the Earth: A Political History of the Space Age* (New York: Basic Books, 1985), p. 4.

14. Stephen Lukashevich, *N. F. Federov* (London: Associated University Presses, 1977), pp. 30, 13, 15, 16, 267.

15. McDougall, *Heavens and Earth,* p. 26; Milton Lehman, *This High Man: The Life of Robert H. Goddard* (New York: Farrar, Straus, 1963), pp. 28, 138, 23; Robert H. Goddard, "Autobiography," in Clarke, *Coming of the Space Age,* pp. 107–8.

16. Tom Crouch, Curator, National Air and Space Museum, Smithsonian Institution, personal correspondence with author, Aug. 29, 1995; McDougall, *Heavens and Earth,* p. 4.

17. David Halberstam, *The Fifties* (New York: Villard, 1993), p. 613; Erik Bergaust, *Wernher von Braun* (Washington, D.C.: National Space Institute, 1976), p. 201. On von Braun's Nazi career, see also Michael J. Neufeld, *The Rockets and the Reich* (New York: Free Press, 1995); Christopher Simpson, *Blowback* (New York: Weidenfeld and Nicolson, 1988).

18. Ernst Stuhlinger, *Von Braun: Crusader for Space* (Malabar, Fla.: Krieger Publishing Company, 1994), pp. 14, 333.

19. Ibid., p. 23.

20. Ibid., p. 332.

21. Author interview with Lucille Johnston, former neighbor of von Braun, April 22, 1993.

22. Lloyd Swenson et al., *This New Ocean: A History of Project Mercury* (Washington, D.C.: NASA, 1966), pp. 29, 523n.; James M. Grimwood, *Project Mercury: A Chronology* (Washington, D.C.: NASA, 1963), pp. 6, 7, 11.

23. Bergaust, *Wernher von Braun,* pp. 282, 285; Anthony M. Springer, "Project Adam: The Army's Man in Space Program," *Quest,* Summer–Fall 1994, pp. 46–47; Buzz Aldrin, *Men from Earth* (New York: Bantam, 1989), pp. 35, 55; "Development Proposal for Project Adam," unpublished manuscript, April 17, 1958, Army Ballistic Missile Agency, Redstone Arsenal, Huntsville, Ala.; "Project Adam: A Chronology," unpublished manuscript, Sept. 11, 1958, ABMA; Dr. Kuettner, memorandum to "All Laboratories," re: "Mercury-Adam Project," Jan. 14, 1959, Huntsville, NASA Historical Documents Collection, NASA Headquarters; Washington, D.C.; telephone conversations between Lt. Col. Walters and S. C. Holmes, Sept. 30, Oct. 1, 1958, NASA HDC; J. B. Medaris to NASA Administrator, Dec. 17, 1958, NASA HDC; John B. Medaris, *Countdown to Decision* (New York: G. P. Putnam's Sons, 1960), p. 116; author interviews with Ernst Stuhlinger, Sept. 23, 1995; with William R. Lucas, Sept. 27, 1995; with John Zierdt, Sept. 23, 1995; with Roger Launius, NASA chief historian, Sept. 28, 1995; T. Keith Glennan, *The Birth of NASA* (Washington, D.C: NASA History Office, 1993), p. 9.

24. Von Braun interviewed in George W. Cornell, "Space Travel Teaches God Much Greater," Huntsville *Times,* July 18, 1969; Charles Reagan Wilson, "American Heavens: Apollo and the Civil Religion," *Journal of Church and State,* vol. 26, p. 217.

25. Wernher von Braun to Mrs. M. J. Kemp, Jan. 3, 1972, NASA HDC; *Christian Century,* Dec. 23, 1959, p. 20.

26. Bergaust, *Wernher von Braun,* p. 177; Wernher von Braun, "Exploration of Space: A Job Calling for International Scientific Cooperation," prepared for the International Aeronautical Federation, Stuttgart, West Germany, 1971, cited in ibid., p. 169.

27. Wernher von Braun, "Responsible Scientific Investigation and Application," unpublished talk presented to the Lutheran Church of America, Philadelphia, Oct. 29, 1976, NASA HDC, p. 74.

28. Ibid., pp. 70, 82; Wernher von Braun, commencement address, St. Louis University, June 3, 1958, quoted in McDougall, *Heavens and Earth,* p. 454; Wernher von Braun to Rev. G. T. Phillips, Dec. 6, 1971, NASA HDC.

29. Stuhlinger, *Von Braun,* p. 331; Wernher von Braun, "Immortality," *This Week*

Magazine, June 24, 1960; Wernher von Braun, "Why I Believe in Immortality," in W. Nichols, ed., *Third Book of Words to Live By* (New York: Simon and Schuster, 1962).

30. Wernher von Braun interviewed by Adon Taft, Miami *Herald,* quoted in Cornell, "Space Travel"; author interview with Lucille Johnston, April 22, 1993; Louis Cassels, "Mysteries of the Universe Confirm Belief in God," *Evening Bulletin* (Philadelphia), June 28, 1969; Wernher von Braun, "What My Religion Means to Me," Huntsville *Times,* March 23, 1968.

31. Stuhlinger, *Von Braun,* p. 273.

32. "Medaris Still as Outspoken as Ever," *Today,* April 16, 1978, p. 6E; Robert Dunnavant, "Military Could Have Carried Off NASA Space Program, Says Medaris," Birmingham *News,* July 1, 1985; see also "John Bruce Medaris," *Ad Astra,* July–Aug. 1991; Medaris, *Countdown to Decision,* passim; Michael Adler, "Two-Star General Becomes Priest," *National Inquirer,* Jan. 10, 1971.

33. Lucille R. Johnston, *Will We Find Our Way? A Space-Age Odyssey* (Atlanta: Cross Roads Books, 1979), passim; Lucille Johnston, *The Space Secret of the Universe* (Birmingham: Roberts and Son, 1969), passim; author interviews with William R. Lucas, July 7, 1993, Sept. 27, 1995.

34. Author interview with Lucas, July 7, 1993; Rodney W. Johnson, quoted in Johnston, *Space Secret,* p. 159; Wernher von Braun to John B. Medaris, Dec. 9, 1971, NASA HDC; "Space Expert Heard in Pulpit," Washington *Post,* Dec. 30, 1968, p. B7.

35. On Chapel of the Astronaut debate, see Congressional hearings on H.R. 11487, Nov. 16, 1971, and H.R. 4545, Sept. 23, 1971, Dec. 2, 1971; J. Bruce Medaris to James C. Fletcher, NASA administrator, June 29, 1973, and Fletcher to Medaris, July 13, 1973, NASA HDC; "Predictions of the Rapture Are Premature," Washington *Post,* Sept. 9, 1989, p. D19; "Rapture That Wasn't Will Be This Year," Washington *Times,* Aug. 25, 1989, p. B5.

36. Author interviews with Jerry Klumas and Tom Henderson, Clear Lake, Texas, Jan. 12 and 13, 1995.

37. Robert E. Bobola, "Examining the Evidence," *Full Gospel Business Men's Voice,* March 1982, pp. 11–15.

38. Author interview with Tom Henderson; "Tom and Judy Henderson Latin American Creation Conferences," May–June 1994; Tom Henderson, "The Social Impact of Evolution," unpublished manuscript, n.d., courtesy of Tom Henderson.

39. Author interview with Jerry Klumas.

40. Michael H. Gorn, *Hugh Dryden's Career in Aviation and Space* (Washington, D.C., NASA History Office, 1966), pp. 11–15; Jo Dibella, "Memorandum re: Dr. Dryden's Church Affiliations," Jan. 10, 1966, NASA HDC; Louis Cassels, "Dr.

Hugh Dryden: Science, Religion Not in Conflict," Washington *Daily News*, July 13, 1963.

41. Hugh Dryden, unpublished sermons, "The Eternal Quest," June 13, 1960; "In the Image of God," Aug. 19, 1951; both in NASA HDC.

42. Hugh Dryden, unpublished sermons, "In the Image of God," Oct. 15, 1961; "Christian Emphasis for Today," Feb. 11, 1951; both in NASA HDC; Hugh Dryden, "The Power of Faith," *Evening Star* (Washington, D.C.), June 1, 1963.

43. Roger D. Launius, "A Western Mormon in Washington, D.C.: James C. Fletcher, NASA and the Final Frontier," *Pacific Historical Review*, vol. 64 (May 1995), p. 217 and passim; interview with Roger Launius, Sept. 28, 1995; Jet Propulsion Laboratory, JPL News Clips, May 8, 1979.

44. "Madalyn Murray Protests Bible Reading from Space," Washington *Star*, Dec. 28, 1968, p. A5; "Court Hears Suit to Bar Space Piety," Washington *Post*, Nov. 25, 1969, p. A8; "Atheist Sues to Prevent Use of Religion in Space," Washington *Post*, Aug. 7, 1969.

45. "Address by Dr. Thomas O. Paine Before the National Press Club," unpublished manuscript, Aug. 6, 1969, Washington, D.C., NASA HDC; "Mail Backs Astronauts on Space Sermons," *New York Times*, Sept. 28, 1969, p. 4.

46. "Atheist Loses Suit to Halt Astronaut's Space Prayers," *New York Times*, Dec. 12, 1969, p. 1; Robert P. Allnut, NASA Assistant Administrator for Legal Affairs, to Rep. Bob Wilson, July 18, 1969; John B. Medaris to James C. Fletcher, June 29, 1973; James C. Fletcher to Roy Ash, Director, Office of Management and Budget, July 13, 1973; John P. Donelly, NASA Office of Public Affairs, to Michael Terrigino, Nov. 2, 1973; all in NASA HDC.

47. O. B. Lloyd, Jr., Director, NASA Office of Public Affairs, to Gerri Madden, June 15, 1972, NASA HDC.

48. "Invitation to Dedication of Space Window at Washington Cathedral," *NASA Headquarters Weekly Bulletin*, July 15, 1974; James C. Fletcher to John D. (no last name given), May 17, 1974; James C. Fletcher to Christopher Kraft, Feb. 5, 1974; James C. Fletcher to Thomas O. Paine, Feb. 5, 1974; Richard Nixon to Thomas O. Paine, Jan. 14, 1974; all in NASA HDC; George Mueller, "Space: The Future of Mankind," *Spaceflight*, March 1985, p. 104.

49. "Memorandum for Record, re: Noah's Ark," NASA ER/Director, Earth Observation Programs, April 5, 1974; ERN/Mr. Centers, "Note to ER/Mr. Stoney," March 6, 1974; W. Stoney to Charles D. Centers, March 6, 1974; "NH-6/Director, Headquarters Administration Division, to Headquarters Employees, re: Scientific Investigations of the Shroud of Turin," June 19, 1979; all in NASA HDC; "St. Christopher Medal in Vanguard," *New York Times*, March 18, 1958; "NASA Engineer Believes Aliens Visited Earth 2600 Years Ago," Los Angeles

Times, Oct. 26, 1973; "Marshall Engineer Develops All-Directional Wheel," *NASA News*, Marshall Space Flight Center, April 8, 1974, NASA HDC.

50. Lewis Mumford, *Pentagon of Power* (New York: Harcourt Brace Jovanovich, 1964), p. 307; "Religion of the Astronauts," unpublished manuscript, n.d., NASA HDC; Brian O'Leary, *The Making of an Ex-Astronaut* (Boston: Houghton Mifflin, 1970), p. 151.

51. "No Room for Agnostics in Space," Los Angeles *Herald Examiner*, May 15, 1963, p. 11.

52. "Presentation of the Astronauts, 1959, Transcript of the Press Conference," unpublished manuscript, pp. 66, 64–65, NASA HDC; "Remarks of Astronaut John Glenn to Congress," *New York Times*, March 1, 1962, p. 15; "Speech by Astronaut John Glenn," *New York Times*, March 2, 1962, p. 18; John H. Glenn, Jr., "Faith Is a Star," *Evening Star* (Washington, D.C.), Dec. 7, 1963.

53. "Presentation of the Astronauts," pp. 64, 65, 66; "Mercury Project Summary," NASA MSC, p. 415, NASA HDC.

54. Wilson, "American Heavens," pp. 221, 220; George W. Cornell, "Astronauts Find God in Space," San Diego *Union*, May 19, 1973; "First Prayer from Space," Apollo 8 transcript, NASA HDC; Frank White, "Space and the Spirit," *New Age Journal*, Jan.–Feb. 1988, p. 40; "Ex-Astronaut Finds Life After Apollo 9," Washington *Post Magazine*, June 4, 1978, p. 5.

55. Russell L. Schweikert, "Earth: Planet 3a of Sol," *Bell Rendezvous*, Spring 1970; Paul Taylor, "Communal Group Sets Sights on Mars," Washington *Post* clipping, n.d., NASA HDC; Don Williams, "How Moon Cast a Spell on Twelve Lives," Fairfax *Journal*, July 20, 1989, p. A2; Robert F. Allnut to Senator Richard Schweiker, July 10, 1969, NASA HDC.

56. Wilson, "American Heavens," pp. 223, 224; "People," *Time*, Nov. 15, 1971, p. 5; "Apollo 11 Mission Commentary," July 23, 1969, NASA HDC; see also Andrew Chaikin, *Man on the Moon* (New York: Viking Penguin, 1994), pp. 204–5.

57. Michael Collins, *Carrying the Fire* (New York: Farrar, Straus and Giroux, 1974), p. 410; "Graham Disputes Nixon on 'Week,'" Washington *Post*, July 26, 1969, p. A10. On the twenty-fifth anniversary of the first moon landing, President Bill Clinton described the Apollo 11 astronauts as "our guides to the wondrous . . . the true handiwork of God" ("Armstrong Calls on Students," Washington *Times*, July 21, 1994, p. A3).

58. Newspaper clippings, NASA HDC: *Auction*, Feb. 1994; Houston *Post*, Dec. 8, 1969; *Christian Science Monitor*, Nov. 8, 1969; Houston *Post*, March 28, 1971; *Florida Today*, Feb. 19, 1995, p. 15; Washington *Post*, Feb. 6, 1971.

59. Chaikin, *Man on the Moon*, p. 443; James Gorman, "Righteous Stuff," *Omni*, May 1984, passim; Cornell, "Astronauts Find God"; Williams, "How Moon Cast

Spell"; see also "James B. Irwin, 61, Ex-Astronaut," obituary, *New York Times*, Aug. 10, 1991, p 26; "Astronaut James Irwin Dies,"obituary, Washington *Post*, Aug. 10, 1991, p. B4; "Former Astronaut Irwin Is in Evangelistic Orbit," St. Louis *Post Dispatch*, Jan. 16, 1973, pp. 10–11; Zeynep Alemdar, "Going to the Mountain," Washington *Post*, Aug. 13, 1986, p. C3; James B. Irwin, *To Rule the Night* (Philadelphia: A. J. Holman Company, 1973), p. 242; James B. Irwin, "Space Explorer's Second Chance," *Full Gospel Business Men's Voice*, March 1982; Eleanor Blau, "Former Astronaut Is on 'High Flight,'" *New York Times*, April 26, 1974; O'Leary, *Making of Ex-Astronaut*, p. 193.

60. Dick Baumbach, "Jesus Biggest Thrill in Astronaut's Life," *Today*, Aug. 11, 1979, p. 10; Henry E. Clements to Christopher Kraft, Oct. 5, 1972, NASA HDC; Charles M. Duke, Jr., "The Adventure Goes On," *Guideposts*, July 1984; Gorman, "Righteous Stuff"; Williams, "How Moon Cast Spell"; Michelle Bearden, "Former Astronauts Explore Spiritual Terrain," St. Petersburg *Times*, Nov. 18, 1989, p. 7E.

61. Jack Lousma, "Nine and a Half Weeks in Space," *Full Gospel Business Men's Voice*, July 1985; Jack Lousma, "Words to Grow On," *Guideposts*, June 1983; *Christian Reader*, July–Aug. 1982; James Warren, "Astronaut Lousma Looks Heavenward for His Guidance," Chicago *Sun-Times*, July 14, 1982, p. 56; George W. Cornell, "Astronauts Find Science, Religion Allied," Phoenix *Gazette*, Dec. 27, 1975, p. A1; Don Lind, talk in Church of the Latter-Day Saints, Lyndonville, Vt., Aug. 16, 1995.

62. Cong. Bill Nelson, letter to constituents, Nov. 1986, NASA HDC: author interviews with Shuttle astronauts Dave Leestma and Joe Tanner, Clear Lake, Texas, Jan. 12, 1995; Robert C. Springer, "Decision," *Full Gospel Business Men's Voice*, July–Aug. 1983.

63. Frank D. Roylance, "Earth: It's Like God Took a Paintbrush," *NASA Current News*, May 2, 1994; Baumgardt, *Johannes Kepler*, p. 34.

CHAPTER 10. THE IMMORTAL MIND: ARTIFICIAL INTELLIGENCE

1. Carola Baumgardt, *Johannes Kepler: Life and Letters* (New York: Philosophical Library, 1951), p. 197.

2. René Descartes to Silhon, May 1637, quoted in Susan Bordo, *The Flight to Objectivity* (Albany: SUNY Press, 1987), p. 26.

3. Ibid., pp. 78, 23, 43, 90, 89.

4. Ibid., pp. 89, 90; Umberto Eco, *The Search for the Perfect Language* (London: Blackwell, 1995), passim; Russell Fraser, *The Language of Adam* (New York: Columbia University Press, 1977), p. 2.

5. Bordo, *Flight to Objectivity*, p. 90.

6. Desmond MacHale, *George Boole: His Life and Work* (Dublin: Boole Press, 1985), pp. 19, 43.

7. George Boole, "The Right Use of Leisure" (1847), quoted in MacHale, *George Boole*, p. 43; MacHale, *George Boole*, p. 69.

8. Ibid., p. 195.

9. George Boole, *An Investigation of the Laws of Thought* (Dover, England: Dover Publications, 1854), pp. 407, 420, 421, 417.

10. MacHale, *George Boole*, pp. 174, 178–79.

11. Sherry Turkle, *The Second Self* (New York: Simon and Schuster, 1984), passim.

12. Andrew Hodges, *Alan Turing: The Enigma of Intelligence* (London: Unwin Paperbacks, 1983), p. 250.

13. Ibid., p. 251; Alan Turing, "Computing Machines and Intelligence," in Edward Feigenbaum, ed., *Computers and Thought* (New York: McGraw-Hill, 1963), pp. 12, 19, 35.

14. Hodges, *Alan Turing*, p. 266; Turing, "Computing Machines and Intelligence," pp. 33, 20.

15. Turing, "Computing Machines and Intelligence," p. 21.

16. Hodges, *Alan Turing*, pp. 520, 512–13.

17. Allen Newell and Herbert A. Simon, *Human Problem Solving* (Englewood Cliffs, N.J.: Prentice-Hall, 1972), p. 881; Douglas D. Noble, "Cockpit Cognition: Education, the Military, and Genetic Engineering," *AI and Society*, no. 3, Fall 1989, pp. 271–96.

18. Robert Wright, *Three Scientists and Their Gods* (New York: Times Books, 1988), p. 31.

19. Ibid.

20. Ibid., p. 45; Pamela McCorduck, *Machines Who Think* (San Francisco: W. H. Freeman, 1979), pp. 346, 351; Eco, *Search for the Perfect Language*, p. 311.

21. Newell and Simon, *Human Problem Solving*, p. 870; Allen Newell and Herbert A. Simon, "Elements of a Theory of Human Problem Solving" (1958), in Herbert A. Simon, ed., *Models of Thought* (New Haven: Yale University Press, 1989), p. 19; Russell, quoted in Herbert A. Simon, *Models of My Life* (New York: Basic Books, 1991), pp. 207, 209, 180–88.

22. McCorduck, *Machines Who Think*, p. 93.

23. Turkle, *Second Self*, p. 264; Jeremy Bernstein, *Science Observed* (New York: Basic Books, 1982), p. 124; Marvin Minsky, *The Society of Mind* (New York: Simon and Schuster, 1985), pp. 322, 323.

24. Marvin Minsky, "Thoughts About Artificial Intelligence," in Raymond Kurzweil, ed., *The Age of Intelligent Machines* (Cambridge, Mass.: MIT Press, 1990), pp. 214, 215, 218.

25. Marvin Minsky, "Steps Toward Artificial Intelligence," in Feigenbaum, ed., *Computers and Thought*, p. 450.

26. McCorduck, *Machines Who Think*, p. 113; Noble, "Cockpit Cognition," pp. 280, 282; "Top Gun and Beyond," "Nova" television program, WGBH (Boston), Jan. 20, 1988; see also H. Sackman, *Computers, Systems Science, and Evolving Society: The Challenge of Man-Machine Systems* (New York: John Wiley, 1967), p. 564; Robert Fano, "The MAC System," in M. A. Sass and W. D. Wilkinson, eds., *Computer Augmentation of Human Reasoning* (Washington, D.C.: Spartan Books, 1965), pp. 131–49.

27. Manfred Clynes and Nathan Kline, "Cyborgs and Space," *Astronautics*, Sept. 1960; see also Chris Habls Gray, *The Cyborg Handbook* (New York: Routledge, 1996).

28. Allucquere Rosanne Stone, "Will the Real Body Please Stand Up," in Michael Benedikt, ed., *Cyberspace: First Steps* (Cambridge, Mass.: MIT Press, 1991), pp. 90, 96.

29. Nicole Stenger, "Mind Is a Leaking Rainbow," in Benedikt, ed., *Cyberspace*, pp. 58, 52.

30. Michael Heim, *The Metaphysics of Virtual Reality* (New York: Oxford University Press, 1993), pp. 95, 104; Michael Heim, "Erotic Ontology of Cyberspace," in Benedikt, ed., *Cyberspace*, pp. 61, 73, 69; Stone, "Will the Real Body," pp. 90, 112.

31. Marcos Novak, "Liquid Architecture in Cyberspace," in Benedikt, ed., *Cyberspace*, p. 241; Michael Benedikt, "Introduction," in Benedikt, ed., *Cyberspace*, pp. 6, 14, 15.

32. Turkle, *Second Self*, pp. 249, 271; McCorduck, *Machines Who Think*, p. 353.

33. Daniel Crevier, *AI: The Tumultuous History of the Search for Artificial Intelligence* (New York: Basic Books, 1993), pp. 278–80.

34. Ibid., pp. 339, 340.

35. Hans Moravec, *Mind Children: The Future of Robot and Human Intelligence* (Cambridge, Mass.: Harvard University Press, 1988), pp. 4, 5, 75, 112, 118; Hans Moravec, *The Age of Mind: Transcending the Human Condition Through Robots*, cited as forthcoming in Roger Penrose, *Shadows of the Mind* (New York: Vintage, 1995).

36. Moravec, *Mind Children*, pp. 121, 122, 123–24.

37. Ibid., p. 4; Crevier, *AI*, p. 339.

38. Steven Levy, "A-Life Nightmare," *Whole Earth Review*, no. 76, Fall 1992, p. 39.

39. Ibid., pp. 39, 40, 41.

40. Wright, *Three Scientists*, pp. 80, 69.

41. McCorduck, *Machines Who Think*, pp. 352, 353, 356; Wright, *Three Scientists*, p. 80; Moravec, *Mind Children*, p. 116.

42. Earl Cox and Gregory Paul, *Beyond Humanity: CyberRevolution and Future Mind* (Cambridge: Charles River Media, 1996), pp. 1–7.

43. Steven Levy, *Artificial Life: The Quest for a New Creation* (New York: Pantheon, 1992), pp. 14, 11.

44. Levy, *Artificial Life*, pp. 34, 37, 40; see also Robert A. Freitas, Jr., and William P. Gilbreath, eds., *Advanced Automation for Space Missions*, NASA Conference Publication 2255 (Springfield, Va.: National Technical Information Services, 1982).

45. Levy, *Artificial Life*, p. 41.

46. Ibid., pp. 36, 113, 95.

47. Ibid., p. 85; Rudy Rucker, *Getting Started*, CA Lab Software, Autodisk, 1989, pp. 17, 18.

48. J. Doyne Farmer and Aletta d'A. Belin, *Artificial Life: The Coming Evolution*, Los Alamos Publication L.A. UR-90, 378, p. 1.

49. Ibid., pp. 22, 23.

50. Levy, "A-Life Nightmare," pp. 36, 46.

51. Ibid., pp. 37, 38, 40.

52. Stefan Helmreich, "Anthropology Inside and Outside the Looking-Glass Worlds of Artificial Life," unpublished manuscript, Department of Anthropology, Stanford University, pp. 6, 9.

53. Ibid., pp. 18, 19.

54. Rucker, *Getting Started*, p. 16; Helmreich, "Anthropology Inside," p. 6.

55. Levy, "A-Life Nightmare," p. 38; Helmreich, "Anthropology Inside," pp. 6, 7, 10, 14.

56. Rucker, *Getting Started*, p. 18; Levy, "A-Life Nightmare," pp. 42, 43; Babbage, *Ninth Bridgewater Treatise*, p. 173.

CHAPTER 11. POWERS OF PERFECTION:
GENETIC ENGINEERING

1. Edward Bellamy, *Looking Backward* (New York: New American Library, 1960), p. 194; J. Doyne Farmer and Aletta d'A. Belin, *Artificial Life: The Coming Evolution*, Los Alamos Publication L.A. UR-90, 378, p. 13.

2. John Cohen, *Human Robotics: Myth and Science* (New York: A. S. Barnes, 1967), pp. 40, 68–69; Margaret Wertheim, *Pythagoras' Trousers* (New York: Times Books, 1995), pp. 155–56; Lewis Mumford, *Pentagon of Power* (New York: Harcourt Brace Jovanovich, 1964), p. 117; Francis Bacon, *The New Atlantis* (New York: Kessinger Publishing Company, 1992), passim.

3. J. D. Bernal, *The World, The Flesh, and the Devil: An Enquiry into the Future of the Three Enemies of the Rational Soul* (Bloomington: Indiana University Press, 1969); see also Edward Yoxen, *The Gene Business* (London: Pan Books, 1983), pp. 33, 41.

4. Bernal, *World, Flesh, and Devil*, pp. 14, 33; J. B. S. Haldane, "The Last Judgement," in J. B. S. Haldane, *Possible Worlds* (London: Chatto and Winders, 1927), cited in Mary Midgley, *Science as Salvation* (London: Routledge, 1992), p. 25.

5. Bernal, *World, Flesh, and Devil*, pp. 37, 61.

6. Ibid., pp. 42, 47.

7. Ibid., pp. 65, 75, 79.

8. Ibid., pp. 66, 45, v.

9. Ibid., p. vi; Arthur Peacocke, *God and the New Biology* (New York: Harper and Row, 1986), p. 60.

10. Jan Sapp, "The Nine Lives of Gregor Mendel," in H. E. LeGrand, ed., *Experimental Inquiries* (Amsterdam: Kluwer Academic Publishers, 1990), p. 10; L. A. Callender, "Gregor Mendel: An Opponent of Descent with Modification," *History of Science*, vol. 26 (1988), p. 41.

11. Gunther S. Stent, *The Coming of the Golden Age* (Garden City, N.Y.: Natural History Press, 1969), p. 7.

12. Horace Judson, *The Eighth Day of Creation* (London: Jonathan Cape, 1979); Weaver, quoted in Lily Kay, *The Molecular Vision of Life* (New York: Oxford University Press, 1993), p. 43; Warren Weaver, *Scene of Change: A Lifetime in American Science* (New York: Charles Scribner's Sons, 1970), quoted in Philip J. Regal, "Biotechnological Jitters: Will They Blow Over?," *Biotechnology Education*, vol. 1 (1989), p. 53.

13. Stent, *Coming of the Golden Age*, p. 35; Ernst Peter Fischer and Carol Lipson, *Thinking About Science: Max Delbrück and the Origins of Molecular Biology* (New York: W. W. Norton, 1988), passim; Peacocke, *God and the New Biology*, p. 57; G. S. Stent, "That Was the Molecular Biology That Was," *Science*, vol. 160 (1968), pp. 390–95.

14. Richard C. Lewontin, "The Dream of the Human Genome," *New York Review of Books*, May 28, 1992, p. 31; see also Fischer and Lipson, *Thinking About Science*, p. 183; Stephen Jay Gould, "'What is Life?' as a Problem in History," in Michael P. Murphy and Luke A. J. O'Neill, eds., *What Is Life? The Next Fifty Years* (Cambridge: Cambridge University Press, 1995), pp. 25–26.

15. Erwin Schrödinger, *What Is Life? The Physical Aspect of the Living Cell* (Cambridge: Cambridge University Press, 1955), pp. 1, 80.

16. Ibid., pp. 1, 61, 88, 86, 79.

17. Ibid., pp. 88, 89.

18. Ibid., pp. 91, 92; on the response of scientific colleagues to his epilogue, see Evelyn Fox Keller, *Refiguring Life* (New York: Columbia University Press, 1995), pp. 76–77; E. F. Keller to author, Jan. 26, 1996.

19. James B. Watson, *The Double Helix* (New York: Atheneum, 1968), pp. 220, 197, 153; Peacocke, *God and the New Biology*, p. 60; Erwin Chargaff, book review in *Perspectives in Biology and Medicine*, vol. 19 (1976), p. 290; Dorothy Nelkin, *The DNA Mystique* (New York: Freeman, 1995).

20. Brenner, quoted in Peacocke, *God and the New Biology*, p. 59.

21. Mumford, *Pentagon of Power*, p. 125; Marshall Nirenberg, "Will Society Be Prepared?," *Science*, vol. 633 (1967), p. 157.

22. Susan Wright, *Molecular Politics* (Chicago: University of Chicago Press, 1994), passim; Burke Zimmerman, *Biofuture: Confronting the Genetic Era* (New York: Plenum Press, 1984), pp. 258–59.

23. Andrew Kimbrell, *The Human Body Shop* (San Francisco: Harper San Francisco, 1993), p. 214.

24. Francis Bacon, quoted in Mumford, *Pentagon of Power*, p. 117.

25. Kimbrell, *Human Body Shop*, pp. 174–87.

26. Ibid., p. 164.

27. Ibid., pp. 169–70.

28. Ibid., p. 171.

29. Ibid., p. 172.

30. Ibid., p. 125; Bernal, *World, Flesh, and Devil*, pp. 73, 76; Kimbrell, *Human Body Shop*, p. 125; Constance Holden, "On the Trail of Genes for IQ," *Science*, vol. 253, p. 1352; Mumford, *Pentagon of Power*, p. 186.

31. Muller, quoted in Evelyn Fox Keller, "Nature, Nurture, and the Human Genome Project," in Daniel J. Kevles and Leroy Hood, eds., *The Code of Codes* (Cambridge, Mass.: Harvard University Press, 1992); Weaver, quoted in Kay, *Molecular Vision*, pp. 283–84, 289, 290; see also H. J. Muller, "Social Biology and Population Improvement," *Nature*, vol. 144 (1939), pp. 521–22.

32. Robert Sinsheimer, "The Prospect of Designed Genetic Change," *Engineering and Science*, vol. 32 (1969), pp. 8–13.

33. Alan Turing, "Computing Machines and Intelligence," in Edward Feigenbaum, ed., *Computers and Thought* (New York: McGraw-Hill, 1963), p. 13; Kimbrell, *Human Body Shop*, pp. 225, 213.

34. Daniel J. Kevles, "Out of Eugenics: The Historical Politics of the Human Genome," in Kevles and Hood, eds., *Code of Codes*, pp. 18–19, vii; Robert L. Sinsheimer, "The Santa Cruz Workshop—May, 1985," *Genomics*, vol. 5 (1989), p. 955.

35. Robert L. Sinsheimer, *The Strands of Life* (Berkeley: University of California Press, 1994), pp. 2, 283, 287.

36. Ibid., p. 3.

37. R. Lewin, "In the Beginning Was the Genome," *New Scientist*, July 2, 1990, p. 36; Kevles, "Out of Eugenics," p. 10.

38. James D. Watson, "A Personal View of the Project," in Kevles and Hood, eds., *Code of Codes*, pp. 164–65; Leroy Hood, "Biology and Medicine in the Twenty-first Century," in ibid., p. 163; Walter Gilbert, "A Vision of the Grail," in ibid., p. 83; "Francis Collins," *Current Biography*, June 1994, p. 9; Gina Kolata, "Unlocking the Secrets of the Genome," *New York Times*, Nov. 30, 1993, p. C1.

39. Janice G. Raymond, *Women as Wombs* (San Francisco: Harper San Francisco, 1993), p. xiii; Chargaff, quoted in ibid., p. xiii; "First Person," NBC, April 4, 1994.

40. Lewontin, "Dream of the Human Genome," p. 31; Roberto Zapperi, *The Pregnant Man* (London: Harwood Academic Publisher, 1991), pp. 3–5.

41. Author interview, Nov. 1995 (this source requested anonymity).

42. Mark Adams, *The Evolution of Theodosius Dobzhansky* (Princeton: Princeton University Press, 1994), passim; Theodosius Dobzhansky, *The Biology of Ultimate Concern* (New York: New American Library, 1967): Peacocke, *God and the New Biology*, pp. 84, 85; J. Robert Nelson, *On the New Frontiers of Genetics and Religion* (Grand Rapids, Mich.: William B. Eerdman, 1994), pp. 21, 16.

43. A. R. Peacocke, *Creation and the World of Science* (Oxford: Clarendon Press, 1979), p. 305; Trinkhaus, quoted in ibid., p. 305; see also Peacocke, *God and the New Biology*, p. 58.

44. Peacocke, *Creation and the World of Science*, pp. 305–6.

45. "Statement of Faith," Membership Application, American Scientific Affiliation; author interviews with Jaydee Hansen, J. Robert Nelson, and Daniel Kevles, Oct., Nov., Dec. 1995; J. Robert Nelson, *Genetics and Religion* (Houston: Institute of Religion, Texas Medical Center, 1995), pp. 171–92.

46. George Liles, "God's Work in the Lab: Geneticist Francis Collins Makes the Case for Faith," *MD Magazine*, March 1992, pp. 43–50; Francis Collins, "Healing Responsibly: The Church and the Human Genome Project," remarks delivered at conference on "The Christian Stake in Genetics," Trinity International University, Deerfield, Ill., July 19, 1996, p. 3.

47. Author interview with Donald Munro, ASA director, Nov. 1995.

48. Ibid.; Hessel Bouma et al., *Christian Faith, Health, and Medical Practice* (Grand Rapids, Mich.: William B. Eerdman, 1989).

49. Bruce R. Reichenbach and V. Elving Anderson, *On Behalf of God: A Christian Ethic for Biology* (Grand Rapids, Mich.: William B. Eerdman, 1995), pp. 58, 178.

50. Ibid., pp. 183, 187, 203.

51. Ibid., pp. 50–51.

52. Nelson, *Genetics and Religion*, passim.

53. C. Thomas Caskey, "Foreword," in ibid., p. ix; author interview with Jaydee Hansen, assistant general secretary, Ministry of God's Creation, for the General Board of Church and Society of the United Methodist Church, and a leading member of the United Methodist Genetic Science Task Force, Nov. 1995; "Special Issue: Genetic Science," *Christian Social Action* (Washington, D.C.: General Board of Church and Society of the United Methodist Church), Jan. 1991; Richard Stone, "Religious Leaders Oppose Patenting Genes and Animals," *Science*, vol. 268 (May 26, 1995), p. 1126.

54. Liles, "God's Work," p. 48.

55. W. French Anderson, "Can We Alter Our Humanness by Genetic Engineering?," unpublished address given at the Conference on Genetics, Religion, and Ethics, Institute of Religion and Ethics, Texas Medical Center, March 13, 1992 (previously presented at conference at the National Cathedral, Washington, D.C., Nov. 1991) (courtesy J. Robert Nelson).

56. Nelson, *Genetics and Religion*, pp. 60, 180; author interview with Sheldon Krimsky, former member of the Recombinant DNA Advisory Committee of the National Institutes of Health, Nov. 1995; "Fact Sheet," Millennium Pharmaceuticals, Inc., Cambridge, Mass., Fall 1993.

57. Daniel Kevles to author, Oct. 23, 1995; Kevles, "Out of Eugenics," p. 36.

CONCLUSION: THE POLITICS OF PERFECTION

1. Robert L. Sinsheimer, *The Strands of Life* (Berkeley, University of California Press, 1994), p. 2.

2. Jacques Le Goff, *Time, Work, and Culture in the Middle Ages* (Chicago: University of Chicago Press, 1980), pp. 80, 186.

3. George Ovitt, *The Restoration of Perfection* (New Brunswick, N.J.: Rutgers University Press, 1986), p. 153; Norman Cohn, *The Pursuit of the Millennium* (Oxford: Oxford University Press, 1961), p. 109; Bernard McGinn, "Apocalyptic Traditions and Spiritual Identity in Thirteenth Century Religious Life," in E. Rozanne Elder, ed., *The Roots of the Modern Christian Tradition* (Kalamazoo, Mich.: Cistercian Publications, 1984), p. 1.

256 / NOTES TO PAGES 203—211

4. Roger Bacon, *The Opus Majus of Roger Bacon* (New York: Russell and Russell, 1962), p. 634.

5. Frances Yates, *The Rosicrucian Enlightenment* (Boulder: Shambala Press, 1978), chap. one.

6. Margaret C. Jacob, *The Cultural Meaning of the Scientific Revolution* (Philadelphia: Temple University Press, 1988), p. 34; James R. Jacob, "By an Orphean Charm," in Phyllis Meade and Margaret C. Jacob, eds., *Politics and Culture in Early Modern Europe* (Cambridge: Cambridge University Press, 1988), pp. 235, 238, 239, 244.

7. J. R. Jacob, "Orphean Charm," pp. 241, 249.

8. Margaret C. Jacob, "Millenarianism and Science in the Late Seventeenth Century," *Journal of the History of Ideas*, vol. 37 (1976), p. 340; see also Margaret Jacob, *Living the Enlightenment* (New York: Oxford University Press, 1991), p. 208; Gertrude Lenzer, *Auguste Comte and Positivism: The Essential Writings* (New York: Harper and Row, 1975), pp. 389, 447.

9. Philip J. Regal, "Scientific Principles for Ecologically-Based Risk Assessment of Transgenic Organisms," *Molecular Ecology*, vol. 3 (1994), p. 5.

10. Lynn White, "The Historical Roots of Our Ecological Crisis," in Lynn White, *Machina ex Deo: Essays in the Dynamism of Western Culture* (Cambridge, Mass.: MIT Press, 1968), p. 89; Philip J. Regal, "Metaphysics in Genetic Engineering: Cryptic Philosophy and Ideology in the 'Science' of Risk Assessment," in Ad Van Dommelen, *Coping with Deliberate Release: The Limits of Risk Assessment* (Amsterdam: Free University of Amsterdam, 1996), p. 25.

11. Mary Midgley, *Science as Salvation* (London: Routledge, 1992), p. 221.

12. Reinhart Maurer, "The Origins of Modern Technology in Millenarianism," in Paul T. Durbin and Friedrich Rapp, eds., *Philosophy and Technology* (Dordrecht: D. Reidel, 1983), p. 265.

13. Cynthia Cockburn, *Machinery of Dominance* (London: Pluto Press, 1985), p. 255; Lewis Mumford, "An Appraisal of Lewis Mumford's *Technics and Civilization*," *Daedalus*, Summer 1959, p. 536.

APPENDIX. A MASCULINE MILLENNIUM:
A NOTE ON TECHNOLOGY AND GENDER

1. Autumn Stanley, *Mothers and Daughters of Invention* (Metuchen, N.J.: Scarecrow Press, 1993), pp. 747, xxxvii.

2. Paola Tabet, "Hands, Tools, Weapons," *Feminist Issues*, vol. 2, no. 2 (Fall 1982), pp. 3–62; Emilie Amt, ed., *Women's Lives in Medieval Europe* (New York: Routledge, 1993), pp. 179, 194, 197; Martha C. Howell, *Women, Production, and Pa-*

triarchy in Late Medieval Cities (Chicago: University of Chicago Press, 1986), pp. 2–5; see also Ivan Illich, *Gender* (New York: Pantheon, 1982), pp. 88–102.

3. Howell, *Women, Production, and Patriarchy*, passim; see also Olive Schreiner, *Woman and Labour* (Cape of Good Hope, South Africa: Frederick A. Stokes, 1911); Ivy Pinchbeck, *Women Workers and the Industrial Revolution* (London: Cass, 1969).

4. Tabet, "Hands, Tools, Weapons"; Jacques Le Goff, *Time, Work, and Culture in the Middle Ages* (Chicago: University of Chicago Press, 1980), p. 186.

5. St. Augustine, *The City of God* (Garden City, N.Y.: Doubleday, 1958), pp. 526–27.

6. Book of Genesis 1:27, 1:26; Gerhart B. Ladner, *The Idea of Reform* (New York: Harper and Row, 1967), pp. 173, 233, 59; Arnold Williams, *The Common Expositor: An Account of the Commentaries on Genesis, 1527–1633* (Chapel Hill: University of North Carolina Press, 1948), p. 26.

7. Tertullian, "Disciplinary, Moral, and Ascetical Works," quoted in Marina Warner, *Alone of All Her Sex* (New York: Alfred A. Knopf, 1976), p. 58.

8. Adela Yerbro Collins, *Crisis and Catharsis: The Power of the Apocalypse* (Philadelphia: Westminster Press, 1984), pp. 127, 129, 131; Kevin Harris, *Sex, Ideology and Religion: The Representation of Women in the Bible* (Totowa, N.J.: Barnes and Noble, 1984), pp. 112–13.

9. Ernst Benz, *Evolution and Christian Hope* (Garden City, N.Y.: Doubleday, 1975), p. 13; on the historical evolution of this clerical world without women, see David F. Noble, *A World Without Women: The Christian Clerical Culture of Western Science* (New York: Alfred A. Knopf, 1992), part two.

10. Erigena, quoted in Georges Duby, *The Knight, The Lady, and the Priest* (New York: Pantheon, 1983), p. 50; ibid., p. 51.

11. Jean Gimpel, *The Medieval Machine: The Industrial Revolution of the Middle Ages* (London: Penguin, 1977), passim; Hugh, quoted in George Ovitt, *The Restoration of Perfection* (New Brunswick, N.J.: Rutgers University Press, 1986), p. 120.

12. Marjorie Reeves, *The Influence of Prophecy in the Later Middle Ages: A Study in Joachimism* (Oxford: Oxford University Press, 1969), pp. 248–50.

13. Roger Bacon, *The Opus Majus of Roger Bacon* (New York: Russell and Russell, 1962), pp. 52, 56.

14. Michael Adas, *Machines as the Measure of Men: Science, Technology and Ideologies of Western Dominance* (Ithaca, N.Y.: Cornell University Press, 1989), pp. 13–14.

15. Charles G. Nauert, Jr., *Agrippa and the Crisis of Renaissance Thought* (Urbana: University of Illinois Press, 1965), pp. 48, 284.

16. Franz Hartmann, *The Life and Doctrines of Philippus Theophrastus* (New York: Theosophical Publishing Company, 1910), pp. 99–101; Wilhelm Waetzgoldt, *Dürer and His Time* (London: Phaidon Press, 1950), p. 207.

17. Steven Ozment, *Protestants: The Birth of a Revolution* (New York: Doubleday, 1992), pp. 151–52; John Milton, *Paradise Lost* (New York: Macmillan, 1993), p. 571 (bk. 10, lines 888–95).

18. Frances Yates, *The Rosicrucian Enlightenment* (Boulder: Shambala Press, 1978), p. 47.

19. Francis Bacon, "De Augmentis," quoted in Williams, *Common Expositor*, p. 81; Francis Bacon, "The Masculine Birth of Time," in Benjamin Farrington, *The Philosophy of Francis Bacon* (Chicago: University of Chicago Press, 1964), pp. 533–54; Francis Bacon, "History of Winds," in Benjamin Farrington, ed., *The Works of Francis Bacon* (Philadelphia: Carey and Hart, 1848), vol. 1, p. 54; Carolyn Merchant, *The Death of Nature* (New York: Harper and Row, 1980), pp. 172–74, 181.

20. Frank E. Manuel, *Freedom from History* (New York: New York University Press, 1971), p. 109; Robert Boyle, "Of the Usefulness of Natural Philosophy," in *Works of the Honorable Robert Boyle* (London, 1772), vol. 2 p. 14; Robert Boyle, *On Seraphic Love: Motives and Incentives to the Love of God* (London: Henry Herrington, 1661); Oldenburg, quoted in Evelyn Fox Keller, *Reflections on Gender and Science* (New Haven: Yale University Press, 1985), p. 52; Sprat, quoted in Londa Schiebinger, *The Mind Has No Sex? Women in the Origins of Modern Science* (Cambridge, Mass.: Harvard University Press, 1989), p. 138.

21. Walter Charleton, quoted in Brian Easlea, *Witch-Hunting, Magic, and the New Philosophy* (Brighton, England: Harvester Press, 1980), p. 242; Joseph Glanvill, *The Vanity of Dogmatizing* (New York: Columbia University Press, 1931), p. 6; Frank E. Manuel, *The Religion of Isaac Newton* (Oxford: Oxford University Press, 1974), pp. 99, 100.

22. *The Constitutions of the Free-Masons* (New York: J. W. Leonard, 1855), pp. 1, 38, 51; Abner Cohen, "The Politics of Ritual Secrecy," *Man*, vol. 6 (Sept. 1977), p. 121.

23. Margaret Jacob, *Living the Enlightenment* (New York: Oxford University Press, 1991), pp. 21, 122, 126–27, 121.

24. Ibid., pp. 135, 139, 125, 122; Margaret Jacob, "Freemasonry and the Utopian Impulse," in Richard H. Popkin, ed., *Millenarianism and Messianism in English Literature and Thought, 1650–1800* (Leiden: E. J. Brill, 1988), p. 141.

25. Nicholas Hans, *New Trends in Education in the Eighteenth Century* (London: Routledge and Kegan Paul, 1951), p. 208.

26. Sally Hacker, "The Culture of Engineering: Women, Workplace and Machine," *Women's Studies International Quarterly*, vol. 4, no. 3 (1981), pp. 341–43.

27. Auguste Comte to John Stuart Mill, Oct. 5, 1843, reprinted in Kenneth Thompson, *Auguste Comte: The Foundation of Sociology* (New York: John Wiley and Sons, 1975).

28. William J. Broad, *Star Warriors: A Penetrating Look into the Lives of the Young Scientists Behind Our Space Age Weaponry* (New York: Simon and Schuster, 1985), p. 25; Robert Jay Lifton and Eric Markusen, *The Genocidal Mentality* (New York: Basic Books, 1988), p. 118; Carol Cohn, "Nuclear Language," *Bulletin of the Atomic Scientists,* June 1987, p. 68.

29. Erik Bergaust, *Wernher von Braun* (Washington, D.C.: National Space Institute, 1976), p. 499; *Historical Data Book IV* (Washington, D.C.: NASA, 1994), p. 104; Ian Mitroff, *The Subjective Side of Science* (Amsterdam: Elsevier, 1974), p. 144; Donald N. Michael et al., "Summary of Proposed Studies on the Implications of Peaceful Space Activities for Human Affairs," report to NASA by Brookings Institution, Dec. 1960.

30. Jean Chesneaux, *The Political and Social Ideas of Jules Verne* (London: Thames and Hudson, 1972), pp. 82, 16; Peter Costello, *Jules Verne, Inventor of Science Fiction* (London: Hodder and Stoughton, 1978), pp. 31, 35.

31. Costello, *Jules Verne*, p. 35; Andrew Martin, *The Knowledge of Ignorance: From Genesis to Jules Verne* (Cambridge: Cambridge University Press, 1985), p. 189.

32. Martin, *Knowledge*, p. 189.

33. Ibid., p. 215.

34. Newell, quoted in Pamela McCorduck, *Machines Who Think* (San Francisco: W. H. Freeman, 1979), p. 122; Steven Levy, *Hackers* (Garden City, N.Y.: Bantam Doubleday, 1994), p. 83.

35. Levy, *Hackers,* p. 84; Allucquere Rosanne Stone, "Will the Real Body Please Stand Up," in Michael Benedikt, ed., *Cyberspace: First Steps* (Cambridge, Mass.: MIT Press, 1991), p. 103; Sherry Turkle, *The Second Self* (New York: Simon and Schuster, 1984), p. 108; Strimpel, quoted in Barbara Kantrowitz, "Men, Women, and Computers," *Newsweek,* May 16, 1994, p. 50.

36. Stefan Helmreich, "Anthropology Inside and Outside the Looking-Glass Worlds of Artificial Life," unpublished manuscript, Department of Anthropology, Stanford University, pp. 37, 19, 20.

37. James B. Watson, *The Double Helix* (New York: Atheneum, 1968), p. 225; Revelation 14:4.

38. Schreiner, *Woman and Labour*, pp. 45–46; Sherwood Anderson, *Perhaps Women* (Mamaroneck, N.Y.: Paul P. Appel, 1970), p. 56.

INDEX

Strimpel, Oliver, 226
Stuhlinger, Ernst, 129
Stukeley, William, 75
Supreme Court, U.S., 134–5
Szilard, Leo, 105, 107, 112, 114

Tanner, Joe, 142
technology:
  coining of term, 93, 223
  escapist fantasies in, 207
  as eschatology, 22, 29
  other-worldly roots of, 9–10
  present enchantment with, 3–4
  relationship between philosophy and,
    49–50, 57
  relationship between religion and, 3–6,
    21
  relationship between science and, 49,
    52, 57–8, 127
  spiritual sterility of, 4
  as useful innovation, 13
telegraph, advent of, 94
Telford, Thomas, 80
Teller, Edward, 112–14, 224
Tenison, Thomas, 68
Tertullian, 213–14
Theophilus, 18–19
Thessalonians, 109
thinking and thought, 144–9
  Boole on, 145–7
  Descartes on, 144–5, 148
  mathematics on, 145–8
  mechanical reproduction of, 148–9
  see also Artificial Intelligence
"Third Essay" (Comte), 83
Thomas, Keith, 45
Thoreau, Henry David, 92
Thurston, Robert, 95–7
Tillotson, John, 68
Tittle, Ernest Fremont, 108–9
Todd, Peter, 171
"To the Number Three" (Boole), 147
transcendence, 11–14, 50–1
  AI and, 151–2, 154, 160, 163–5
  A-Life and, 170
  engineering and, 80, 82
  genetic engineering and, 174, 180, 189
  millenarianism and, 22, 44–5, 50,
    53–4, 63–5, 67, 71, 84, 104–5, 107,
    111
  and politics of perfection, 206–8
  reform of man to image and likeness of
    God in, 15
  relationship between technology and, 9,
    11–17, 22, 28, 53, 68, 111, 114,
    133, 208
  in scientific revolution in seventeenth
    century, 63–5, 67

  space exploration and, 114, 120, 128,
    133, 139
  and U.S. as new Eden, 92, 94, 97
  women and, 209–10, 212, 228
Treatise on the Millennium (Hopkins), 91
trinitarianism, 10
Trinity test, 106–7
Tsiolkovsky, Konstantin, 120–2
Turing, Alan, 149–53, 188
Turkle, Sherry, 149, 160
Tyndale, William, 43–4

Ulam, Stanislaw, 114, 165–6
Unitarianism, 70, 145–6
United States:
  engineering in, 82, 93, 95–7
  Freemasonry in, 76, 78–9, 88, 93, 98
  millenarianism in, 88–91, 98–9, 103
  as new Eden, 88–100, 103
  Protestantism in, 89–90
  religious revivalism in, 90–3
  space exploration and, 120–38
useful arts, 15
  Freemasonry and, 73–5, 77–9
  millenarianism and, 21–2, 26–9, 31,
    39–42, 44–6, 48–55, 57, 60–3, 67,
    69, 71, 85–6, 90
  and politics of perfection, 202, 205
  relationship between perfectionism and,
    21
  in Renaissance, 35–7
  in scientific revolution in seventeenth
    century, 57, 60–3, 67
  and U.S. as new Eden, 88, 90, 92–3, 96
  women and, 209–12, 214–23
  see also mechanical arts
"Usefulness of Natural Philosophy"
  (Boyle), 58, 60
utilitarianism:
  AI and, 159
  engineering and, 82
  Freemasonry and, 77
  millenarianism and, 48–50, 53, 57, 59,
    62–3, 65, 71
  in scientific revolution in seventeenth
    century, 57–9, 62–3, 65
  and U.S. as new Eden, 93, 97
utopianism:
  atomic weapons and, 105
  Freemasonry and, 77
  genetic engineering and, 196–7
  millenarianism and, 38–42, 53–4, 57,
    77
  space exploration and, 121
  and U.S. as new Eden, 90–1, 94–5,
    97–100, 103
  women and, 220, 226
  see also paradise

A NOTE ON THE TYPE

The text of this book was set in Sabon, a typeface designed by Jan Tschi-
chold (1902–1974), the well-known German typographer. Based loosely
on the original designs by Claude Garamond (c. 1480–1561), Sabon is
unique in that it was explicitly designed for hot-metal composition on
both the Monotype and Linotype machines as well as for filmsetting. Designed in 1966 in Frankfurt, Sabon was named for the famous Lyons
punch cutter Jacques Sabon, who is thought to have brought some of
Garamond's matrices to Frankfurt.

Test Your Own IQ contains a collection of 10 tests, each with 40 IQ-type questions designed to develop your problem-solving abilities. These tests are intended to amuse and instruct. They should be considered only a rough guide for determining IQ.